MARCO BASTOS

BREXIT, TWEETED

Polarization and Social Media Manipulation

First published in Great Britain in 2024 by

Bristol University Press
University of Bristol
1–9 Old Park Hill
Bristol
BS2 8BB
UK
t: +44 (0)117 374 6645
e: bup-info@bristol.ac.uk

Details of international sales and distribution partners are available at
bristoluniversitypress.co.uk

© Bristol University Press 2024

British Library Cataloguing in Publication Data
A catalogue record for this book is available from the British Library

ISBN 978-1-5292-2449-8 hardcover
ISBN 978-1-5292-2450-4 ePub
ISBN 978-1-5292-2451-1 ePdf

The right of Marco Bastos to be identified as author of this work has been asserted
by him in accordance with the Copyright, Designs and Patents Act 1988.

All rights reserved: no part of this publication may be reproduced, stored in
a retrieval system, or transmitted in any form or by any means, electronic,
mechanical, photocopying, recording, or otherwise without the prior permission
of Bristol University Press.

Every reasonable effort has been made to obtain permission to reproduce copyrighted
material. If, however, anyone knows of an oversight, please contact the publisher.

The statements and opinions contained within this publication are solely those
of the author and not of the University of Bristol or Bristol University Press.
The University of Bristol and Bristol University Press disclaim responsibility
for any injury to persons or property resulting from any material published in
this publication.

Bristol University Press works to counter discrimination on
grounds of gender, race, disability, age and sexuality.

Cover design: Lyn Davies Design
Front cover image: Alamy/Jeremy Sutton-Hibbert
Bristol University Press uses environmentally responsible
print partners.
Printed and bound in Great Britain by CPI Group (UK)
Ltd, Croydon, CR0 4YY

Contents

List of Figures and Tables iv
About the Author vi
Acknowledgements vii

| one | Introduction | 1 |
| two | Fifty Million Brexit Tweets | 7 |

PART I The Great Upset
| three | Political Realignment | 27 |
| four | Nationalism and Populism | 34 |

PART II Bots Talking to Bots
| five | Polarization | 49 |
| six | Bots and Trolls | 69 |

PART III Troll Farms Offshore
| seven | Information Warfare | 91 |
| eight | Social Media Manipulation | 104 |

PART IV Politics Erased
| nine | The Politics of Deletion | 117 |
| ten | Accountability of Social Platforms | 136 |

| eleven | Conclusion | 148 |

References 160
Index 183

List of Figures and Tables

Figures

2.1	Brexit tweets posted by UK users	9
2.2	Brexit tweets posted by county (sample of 220,000 messages)	11
3.1	Ideological coordinates of British public opinion and political value space	33
4.1	Area under the curve calculated from train and test data sets for the Economism-Populism and Globalism-Nationalism ideological pairs	41
4.2	Ideological value space calculated from Twitter messages	43
4.3	Mean score of (a) Globalism-Nationalism and (b) Economism-Populism for each parliamentary constituency, and (c) the results of the referendum	44
5.1	(a) Cumulative distribution function of in-bubble (echo chambers) and non-bubble (out- and cross-bubble) communication; (b) histogram of distance travelled by messages between sender and receiver in 50-km bins	60
5.2	(a) Geographic coverage of echo chambers (in-bubble) with number of vertices and edges in each subgraph; (b) central point of diffusion of the Leave campaign in the English Midlands, the North, and the East	62

5.3	Distances covered by interactions across in-bubble, out-bubble, and cross-bubble for the Leave and Remain campaigns, with Kolmogorov-Smirnov test statistic and p-value	64
6.1	Two-tiered botnet, with bots specialized in retweeting active users and bots dedicated to retweeting other bots	78
6.2	Activity of Brexit Bots in the weeks leading up to the UK EU membership referendum	81
6.3	(a) Time-to-cascade and (b) mean cascade time for active users and Twitterbots	84
6.4	Large cascades ($S>506$) from user to user, bot to bot, and user to bot	85
9.1	Tweet and user account decay	126
9.2	Percentage of deleted user accounts and tweets	130
9.3	Deleted versus existing tweets about Brexit from April 2016 to October 2019	133
9.4	Time series of existing and deleted tweets with anomalies identified by the S-H-ESD algorithm	134

Table

7.1	Weblinks tweeted by the Brexit Bots	102

About the Author

Marco Bastos is the University College Dublin Ad Astra Fellow at the School of Information and Communication Studies and Senior Lecturer in Media and Communication in the Department of Media, Culture and Creative Industries at City, University of London. He has held research positions at the University of California at Davis, Duke University, University of São Paulo, and the University of Frankfurt. His research leverages computational methods and network science to explore the intersection of communication and critical data studies. He is the author of *Spatializing Social Media: Social Networks Online and Offline* (Routledge, 2021) and his work has featured in media outlets such as *BBC*, *New York Times*, *Guardian*, *Washington Post*, and *BuzzFeed*.

Acknowledgements

This book is dedicated to my daughter Sofia, who remains unimpressed with all things digital and who in her own way assisted and held back this project in an idiosyncratic and invigorating way. I am also thankful to my colleagues and co-authors who collaborated in this project and without whom it would not have been possible. Special thanks to colleagues who worked on the studies that informed this book, including Andrea Baronchelli, Johan Farkas, Fabio Goveia, Naoise McNally, Dan Mercea, Michael Simeone, Marc Tuters, Otávio Vinhas, and Shawn Walker.

The research underpinning this book was supported by Google, Inc. research program #270353197 *The Visual Frames of Social Media Propaganda* and by Twitter, Inc. research grant 50069SS *The Brexit Value Space and the Geography of Online Echo Chambers*. The author also acknowledges support from the University College Dublin research grant #64927.

ONE

Introduction

The book explores a data set of 45,476,692 tweets posted by 264,766 British users, a cohort we identified as the 'British Twitter Monthly Active Userbase'. The data include posts from April 2016 until the end of January 2021 and were collected in real time and on a rolling basis. The first entries in the database start with the official referendum campaign and the last entries end with the United Kingdom's departure from the European Union (EU) on 31 December 2020. The unique and comprehensive nature of the data underpinning this book offers a valuable perspective on the Brexit debate that emerged on Twitter. We hope we can successfully communicate the many puzzling and some paradoxical findings drawn from a large, long-term, and rigorous data set of tweets. Given the contentious nature of the Brexit debate, and the disinformation onslaught on social media observed in the period, much of the book is about social media disinformation, manipulation, and information warfare.

We take stock of emerging trends in the data pointing towards epochal changes in partisan politics, including the political realignment towards nationalist and populist values, but also broader societal changes feeding into polarization and echo-chamber communication. The book also provides an account of how media manipulation emerged to national and then international attention in the run up to the referendum campaign and in the wake of the Cambridge Analytica data scandal, including a detailed account of techniques employed to interfere and potentially distort the public discussion. The book closes with an analysis of the precarity and ephemerality of social media register, as nearly one-third of messages tweeted about Brexit in the preceding years would ultimately disappear from the public domain.

In contrast to much social media research, this book presents a deep and long-lasting study of public debate on Twitter during the Brexit campaign in 2016 and its subsequent negotiations between the UK and EU. It builds on and synthesizes a large number of existing research studies from a research project carried out at City, University of London and the University College Dublin. The book relies on a series of research papers published with my team of collaborators, including Andrea Baronchelli and Dan Mercea at City, University London, Johan Farkas at Malmö University, and Shawn Walker and Michael Simeone at Arizona State University. These studies are synthesized and made accessible to a non-technical, academic audience.

Our expectation, or should we say our hope, is that the robust and comprehensive data set upon which this study builds may offer solid ground to discuss what remains a rather contentious episode in British and international politics. We have sought to ensure the analyses presented in this book, and the conclusions drawn from it, do not result from speculative comments and accusations, but from rigorous evidence drawn from the wide-ranging set of analyses performed for this project. At times, however, there are simply too many data points linking specific actors – be these affiliated with the Leave or the Remain campaign – with manipulative campaigning. In such cases we believe it was necessary to document and comment on this body of circumstantial evidence.

The book may prove particularly valuable for those interested in the study of disinformation and polarization on social media. It may also prove contentious, as the UK EU membership referendum continues to be a controversial topic and the content of this book may be politicized, owing to its findings and the large data set compiled for the multiple studies underpinning the project. This is particularly the case for the second half of the book, which delves into questions about the social media manipulation and information warfare tactics employed by partisans in the Brexit debate on Twitter. In many instances, we sought to remove indications of particular campaigners who may have driven such tactics, as many of these actors and campaigners are still in politics, and some still occupy prominent political positions.

Twitter, of course, no longer exists having been absorbed and succeeded by X following the contentious acquisition of the social media platform by X Corp, a parent company established by Elon Musk in 2023 as the successor to Twitter, Inc. Elon Musk's stated reasons for purchasing the social media platform intersect with many of the key issues discussed in this book, including bots and automation, but also content moderation, freedom of speech, and the use of social media to spread misinformation and disinformation. While we are well aware that Twitter is now referred to as X, this project was concluded prior to the acquisition of the social media platform by Elon Musk. As such, we decided to keep the terminology and the name of the company consistent with the discussion and the scientific research on social media that owed much to Twitter and its generous stance to data access, a stance that unfortunately changed dramatically with the acquisition and rebranding of the company by Elon Musk. In view of that, we refer to the company as Twitter instead of talking about 'X, formerly known as Twitter'.

The intention of this project is to offer a roadmap to the upheaval caused by social media propaganda. It achieves that by tracing the onset of the Leave and Remain campaigns and the ensuing partisan societal polarization that was reflected on Twitter. The chapters that follow this introduction set the scene for further analysis by revisiting the historical and ideological context against which the Brexit debate unfolded. For this portion of the book, we sought to strike a balance between introducing sufficient background information and the context underpinning the Twitter analysis (outlining, for example, the shifting attitudes towards the EU in British politics from Thatcher to Johnson), after which the book moves swiftly to the core concerns of this project. These first chapters are, therefore, largely intended to unpack the data and offer essential background information, including the methodological framework informing this project.

As such, Chapter Two offers a detailed account of the database underpinning this book. Chapters Three and Four are dedicated to the uptake in populist and nationalist sentiments that were pivotal to the Leave campaign, while Chapter Five traces the formation of echo chambers across the partisan divide. Chapter Six is dedicated to the identification of a large botnet that tweeted the campaign and whose study was the object of considerable press interest. Chapter Seven delves into the work of the Internet Research Agency, a Russian 'troll factory' that operated in the Brexit campaign. Chapter Eight offers a critique to current mitigation strategies to offset disinformation on social media platforms, largely centred around fact-checking, while Chapters Nine and Ten unpack the challenges in performing forensic analysis of social media disinformation given the ephemeral dimensions of the data.

The later chapters work through key concerns related to social media propaganda during the Brexit campaign. Some of these concerns are also applicable to related but perhaps less directly comparable events (for example, contentious elections and highly polarized political debates). Building on previously published studies from this project, the book presents a number of consequential observations about the role of bots and trolls, the use of explicit information warfare and opinion manipulation strategies, and – perhaps most surprising of all – the gradual disappearance of trace evidence for such interference on Twitter, either as a result of the 'natural' decay of such data, or as a result of deliberate attempts of those behind the campaign to cover their tracks. We expect the results presented in this portion of the book to offer important lessons for social media research, platform providers, and regulatory bodies more generally.

Hybrid warfare in a high-choice media environment requires propagandists to work across multiple channels, and the Brexit campaign was one such event that effectively saturated our hybrid media system. Twitter was, of course, just one of the many platforms leveraged by both sides of the campaign, but the implementation of a similarly rigorous data collection pipeline for other social media platforms, where the Brexit psychodrama also unfolded, was unfortunately not feasible. This remains virtually impossible under current data access regimes (and ethically problematic given the considerably more private nature of user activities on platforms such as Facebook and WhatsApp). The data informing this book were, therefore, collected from Twitter, but the context of partisan polarization in a high-choice media environment means that much of the content explored in this book appeared across the broader universe of social media platforms.

The conclusions we draw from the database, however, are explicitly and necessarily limited to the public debate that unfolded on Twitter, and therefore unable to address similar or related activities on other social media platforms such as Facebook with its enormous userbase. Even though smaller, Twitter is a particularly influential platform in public political debate, in the UK and beyond. Its smaller userbase is disproportionately influential as it is used widely by politicians, journalists, activists, and other key political stakeholders, and tweets from such actors are often featured in mainstream media reporting. Indeed, the insights gleaned from this project highlight the platform's embeddedness in a wider and more diverse, online and offline, media environment that is densely interconnected and supports information flows across as well as within platforms.

Finally, while the governance structures overseeing social media platforms are not designed to take into account the diverse affordances of competing social media services, regulation of social platforms needs to be – at least to a certain extent – platform specific. The Cambridge Analytica scandal, for instance, was very much a Facebook scandal that resulted in Facebook's abrupt closure of data access for critical research. Twitter, on the other hand, was fundamentally a public platform relatively amiable to independent research and external oversight.

We hope these insights are clearly unpacked in the following chapters, particularly the second portion of the book that explores the multiple vantage points afforded by the data curated for this project, as well as the constraints in exploring Twitter data.

TWO

Fifty Million Brexit Tweets

This chapter presents the manifold methodological challenges associated with large-scale social media data collection, the creation of a database of monthly active Twitter users in Britain, and the process of triangulating geographic information to determine the user location. This includes the processing of geolocation data and the mapping of users onto electoral districts. We also considered the demographic biases of Twitter, the curation of the database, and the process of identifying political affiliation based on campaign advocacy.

British Twitter Monthly Active Userbase

The complete database curated for this project is at least one order of magnitude larger than the data set we explore in this book. More precisely, the monitoring and archiving of Brexit-related tweets between 2016 and 2021 yielded around 1 billion tweets, but this database includes messages posted by users anywhere in the world. For the purposes of this book, we focus on messages posted by users based in the UK, and more specifically by users whose location we could pinpoint to one of the UK's council wards, local authority districts, or alternatively to borough, district, and council areas. This parameter reduces the database to a manageable, albeit still rather large, set of 50 million messages posted by users whose location can be identified with reasonable precision up to postcode level in the UK.

We relied on the estimates of Rose and McGrory (2016), which put the Twitter userbase in the UK at 12 million during the referendum, to approximate the number of monthly active users (MAU) as the relevant metric to estimate the British userbase (users who logged on at least once in the last month). To this end, we estimated that the British Twitter Monthly Active Userbase (BTMAU) would comprise approximately 3.6 million users. The BTMAU database was collated by identifying users who posted content about Brexit and whose location was pinpointed to the UK, and continuously repeating this process daily to identify more users matching the criteria for inclusion in the BTMAU database on a rolling basis. A dedicated server was configured to continuously retrieve, analyse, and to map messages to postcode level, thus allowing for the detection of the location of users and whether they were active on a monthly basis.

The 50 million messages in the database were posted by around 250,000 British users who were active on Twitter every month. This is the core userbase we refer to as the BTMAU, a group of users who posted about Brexit during or after the official campaign and that we monitored in the years that followed the vote. As we discuss in this book, there are important challenges in mapping social media data onto electoral districts due to the hierarchical subdivision of UK local government into sub-authorities and enumeration districts that only exist in portions of the UK territory. Upon overcoming these challenges, we generated a database of 45 million tweets, with a high average number of tweets per user ($\bar{x} = 171$ and $\tilde{x} = 15$, max = 85,156) due to the active tracking of this cohort of users. Figure 2.1 shows the tweet activity for the BTMAU cohort from April 2016 to January 2021.

By drawing on this large, relatively stable, and comprehensive dataset of Brexit-related public Twitter activity over several years we hope to provide a rigorous overview of how the debate unfolded on social media. Naturally many of the following chapters are dedicated to the official UK EU membership referendum campaign (that is, the weeks that

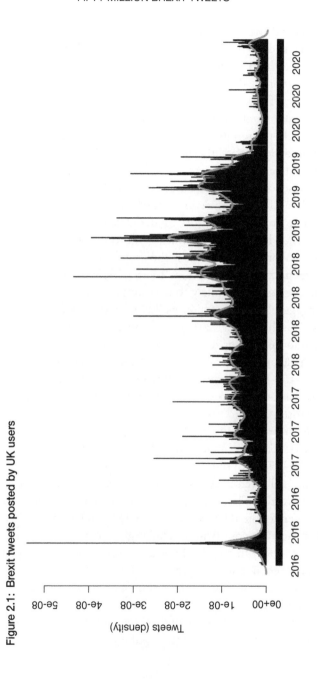

Figure 2.1: Brexit tweets posted by UK users

preceded the vote), but in the second part of the book we unpack the 45 million tweets posted by 265,000 accounts termed the BTMAU. This dataset is a considerable departure from typical, shorter term, and perhaps more cross-sectional studies of Brexit-related Twitter activity that build on datasets of tweets containing the keyword 'Brexit' over a few weeks. As such, the results reported in this book build on a consistent approach to data gathering over time and through a period of several years, thereby enabling the project to provide a robust analysis of patterns and trends in public debate on Twitter, and to offer a unique contribution to the Brexit post-mortem.

The top ten accounts in the database posted a total of 660,777 Brexit-related tweets. One-third of these accounts is no longer available because the user has either been removed, suspended, or blocked from Twitter. Twitter's demographic is, on average, more likely to be younger, urban, and politically engaged; but also white, well educated, and wealthier (Pew Research Center, 2013). A rudimentary classifier for the gender of Twitter accounts identifies 61 per cent of BTMAU as male and 36 per cent as female. This breakdown does not veer too far off industry estimates, which project the female userbase on Twitter at 43 per cent and males at 56 per cent (Statista, 2021).

More surprising is the geographic distribution of the BTMAU, which appears to reflect the population distribution in the UK, with the BTMAU largely based in England (84 per cent), followed by Scotland (10 per cent), Wales (4 per cent), and Northern Ireland (2 per cent). London alone is home to 19 per cent of the BTMAU, followed by Manchester (4 per cent), and then Glasgow, Bristol, Edinburgh, Leeds, Birmingham, and Liverpool with around 2 per cent each. These urban centres alone account for 35 per cent of the BTMAU. Figure 2.2 shows the geographic distribution of Brexit tweets per county in the UK.

A considerable portion of these posts has been removed from Twitter since the time of their publishing, a reflection on the

Figure 2.2: Brexit tweets posted by county (sample of 220,000 messages)

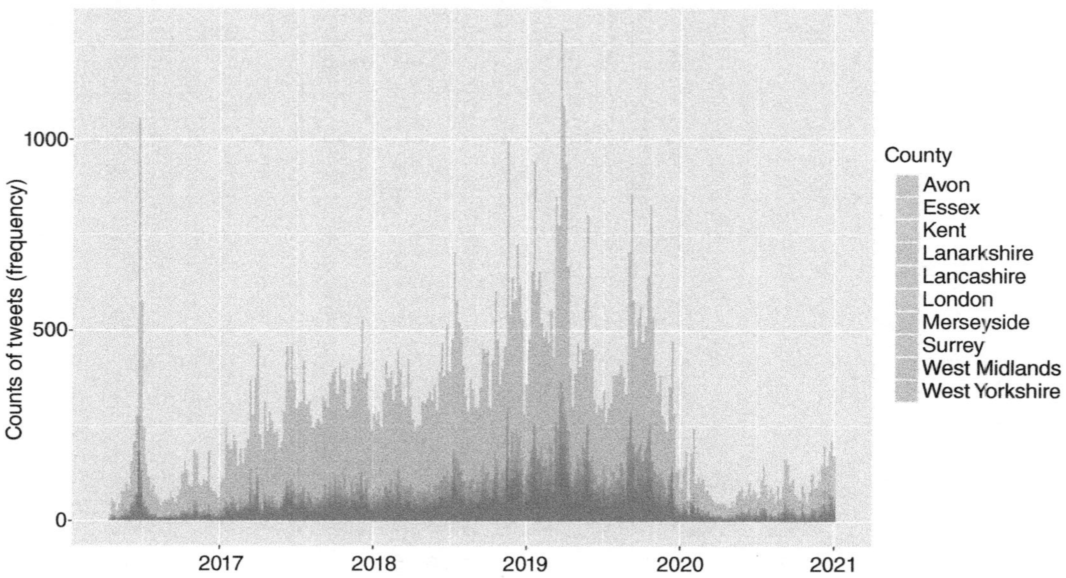
Counts of tweets (frequency)
1000
500
0
2017
2018
2019
2020
2021
County
Avon
Essex
Kent
Lanarkshire
Lancashire
London
Merseyside
Surrey
West Midlands
West Yorkshire

Note: Metropolitan areas account for 35 per cent of the content posted by the British Twitter Monthly Active Userbase.

ephemerality of social media posts that is at odds of the lasting impact of the debate.

Data collection pipeline

The first part of this project was focused on collecting tweets associated with the UK EU membership referendum that ran in the first half of 2016. We relied on the Twitter Streaming and REST Application Programming Interfaces (APIs) to collect nearly 10 million tweets using a set of keywords and hashtags, including relatively (at that time) neutral tags such as brexit, referendum, inorout, and euref, but also messages that used hashtags clearly aligned with the Leave campaign: voteleave, leaveeu, takecontrol, no2eu, betteroffout, voteout, britainout, beleave, iwantout, and loveeuropeleaveeu; and hashtags clearly aligned with the Remain campaign: strongerin, leadnotleave, votein, voteremain, moreincommon, yes2eu, yestoeu, betteroffin, ukineu, and lovenotleave. Vocal hashtags supporting the campaigns were leveraged to identify messages advocating each side of the referendum: The Vote Leave or Vote Remain campaigns.

Reports on trending and the most commonly used hashtags outside the search criteria were generated daily, inspected on a rolling basis, and added to the pool of terms whenever they proved to be relevant. This approach allowed for the expansion of the initial search criteria by consolidating the set of hashtags recurrently tweeted by users. The queried hashtags were parsed across multiple pools to avoid API filtering. Queries that exceeded the one per cent threshold were parsed across separate queries, cumulatively requiring a combination of 12 independent calls to the Streaming API. For the analysis of the referendum, we removed messages tweeted before 15 April 2016, the starting date of the official campaign period, and 24 June 2016, the end of the referendum campaign. Messages flagged as likely to have been tweeted by bots were also removed in the course of the project (Bastos and Mercea,

2019). The resulting dataset includes campaign-aligned hashtagged tweets that we leveraged to build a classifier to identify affiliation to each side of the referendum: Vote Leave or Vote Remain campaigns.

After collecting data from the Streaming API on a rolling basis, we set up a server to regularly query the Twitter REST API and retrieve the profile of users that tweeted the referendum. This automated process managed to retrieve 95 per cent of the user profiles that appeared in the data. Profile information, along with information tweeted by the users, was pivotal to identifying the location of the userbase. We triangulated information from geocoded tweets (subsequently reverse-geocoded), locations identified in their user profile (then geocoded), and information that appeared in their tweets. The triangulation prioritizes the signal with higher precision, hence geocoded information is preferred if present. When not available, we look at the location field in users' profiles and geocode that location. If neither source of information is available, we check for information in their tweets, but only in cases where the *place_id* field of the API response returns relevant information. HERE provided the API used to geocode and reverse geocode geographic location. As the API provides attribute-level information about the match quality, we removed API responses with a MatchCity score > .9 and whose field MatchType of pointAddress failed to pinpoint the location on the map (HERE, 2013). Even with this extensive triangulation, a considerable portion of user locations could be identified only to city or postcode level. This rigorous and iterative process to identify the location of users was only feasible up to 2020, when Twitter implemented a policy change to prevent the sharing of geocoordinates (Kruspe et al, 2021).

Once the location of users was identified, we relied on the longitude and latitude values to calculate the Euclidean distance (in kilometres) covered by the sender and receiver of @-mentions and retweets. This allowed us to examine whether Leave and Remain interactions are predominantly within neighbouring in-bubbles or geographically proximate echo chambers – that is, within a 50-km radius expanded in 50-km increments up to 900 km – which is the maximum straight-line distance between two geographical points in the United Kingdom (from Land's End to John o' Groats). We used the canonical mean equatorial radius (6378.145 km or 2.092567257e7 ft) for the Earth's radius, R, which means the calculation was not mathematically precise due to this inaccurate estimate. Despite this perennial limitation, we believe the calculation is adequate as mathematical precision is of lesser importance when analysing data whose geographic accuracy is limited to postcode level. We repeat the process for each tweet, thus identifying the account being @-mentioned or retweeted and calculating the distance (in kilometres) between sender and receiver. Finally, differences in distance were analysed with a series of statistical tests, including Chi-square and Kolmogorov–Smirnov. For the Chi-squared tests, we rejected the null hypothesis of the independence assumption if the p-value of

$$x^2 = \sum_{i,j} \frac{\left(f_{i,j} - e_{i,j}\right)^2}{e_{i,j}}$$

was less than the given significance level α.

As a result of these filtering procedures, a considerable portion of the locations in our dataset could be identified only to city or postcode level. Nonetheless, we succeeded at identifying the geographic location of 60 per cent of users that tweeted the referendum, a cohort that forms our population of interest. The universe of British users that tweeted the referendum is, however, at least one order of magnitude smaller than the universe of users that appeared in our data collection pipeline, a figure that speaks to the international profile of the event but also to the cross-border dimensions of political campaigning and deliberation in the early 21st century. Upon identifying the location of users, we removed user accounts located outside the United Kingdom or whose location we could not identify up to postcode level. This further reduces our dataset to yet another order of magnitude, so that the 45 million messages curated for this project stem from at least one billion tweets collected in relation to the Brexit saga.

The post-Brexit data explored in the course of this project were derived by querying a database of 100 million Brexit-related tweets, starting in the period leading up to the referendum campaign and ending in October 2019. This database includes 43 months of Brexit-related messages. For the analyses discussed in Chapter Nine, we took a sample without replacement of 50,000 tweets per month, an approach similar to the constructed week sampling employed in journalism studies that maximizes generalizability beyond consecutive days and is suitable for estimating content for a six-month period or longer (Riffe et al, 1993). The monthly sampling approach returned a data set of 2,150,000 tweets (of which 1,404,704 are retweets), subsequently rehydrated to estimate the fraction of deleted tweets and user accounts in the data for each of the 43 months that followed the referendum campaign.

A computer script was written to programmatically query Twitter REST API for user accounts and tweet IDs (rehydration). These steps allowed us to calculate the tweet decay coefficient for Twitter data sets at scale. We relied on this program to validate the results on a range of hashtags with several thousand tweets posted in the same period of the official campaign of the UK EU membership referendum starting in the first half of 2016. For each hashtag, we take a random sample without replacement of 1,000 tweets in the data, and query Twitter API to verify if the message is still available and whether the account (by user ID) that sourced the content remains active.

We also implemented control mechanisms to identify anomalies in the data collection pipeline. Identifying temporal trends in the presence of anomalies is a non-trivial task for anomaly detection, so we relied on the Seasonal Hybrid Extreme Studentized Deviate (S-H-ESD) algorithm to discover statistically meaningful anomalies in the input time series of deleted and active tweets (Vallis et al, 2014). S-H-ESD employs time series decomposition and Generalized Extreme Studentized Deviate to test for meaningful anomalies in temporal data with inherent seasonal and trend components, such as timestamped social network transaction data. This approach builds upon the Generalized ESD test for detecting outliers introduced by Rosner (1983) to identify global and local anomalies. The algorithm supports long time series such as the one explored in this study, and employs piecewise approximation to identify both positive and negative anomalies in input time series (point-in-time increase versus decrease in tweets), which is important as we are interested both in the upsurge and decline of the fraction of tweets deleted in the observed time period.

Finally, there are important ethical considerations in curating such a large database created from data collected via the publicly accessible Twitter Streaming and REST APIs. Although the information collected is public, there are important ethical issues associated with harvesting public Twitter accounts (Zimmer, 2010). Twitter profiles set to private are not present in our pool of users and no private information was examined in the analysis. While we have looked to preserve users' rights and interests, we occasionally decided to disclose the Twitter handles of accounts identified as bots whenever there was a reasonable level of confidence that we were dealing with trolls or Twitterbots, to which ethical considerations of privacy are immaterial. We also considered the potential sensitivity of some of the tweets analysed for this project. These concerns were restricted to content posted by real users; in the case of the Brexit Bots, we conclude that anonymizing the seeding accounts would impinge on our ability to understand the scope of the botnet and the strategies adopted by botmasters.

We also considered the ethical obligation not to display deleted tweets. While Twitter is a public, self-publishing microblogging platform, with the published data being also public, the reality is of course more nuanced. Even though the data are public, users may hold a reasonable expectation of relative privacy to content they have posted or shared with a small number of contacts. Indeed, Twitter Privacy Policy states its services are public and that private accounts are removed from data streamed through Twitter's Streaming API (Twitter, 2018c), but public trace data can still pose tangible risks to the subjectivity of individual users. In other words, users might hold a reasonable expectation that the publicness of their activities will not infringe on their privacy or make them vulnerable to unintended scrutiny and even abuse (Metcalf and Crawford, 2016). In our case, we quoted users positioned at the centre of the Brexit Botnet when posts surpassed 500 retweets, thus avoiding the risk of exposing potentially unnoticed content. We also removed the username that authored the content and identified retweets as exchanges between active users and/or bots. For this cohort of accounts, and regardless of the level of automation involved, we expected users to have a clear sense of the publicness of their quoted posts.

There are other challenges beyond quoting retweets verbatim that cannot be offset by simply inspecting individual tweets or performing the bulk of the analyses on aggregate data. Many of the Twitter handles flagged as Brexit Bots included valuable information to the story we sought to tell, including political slogans, party or campaign affiliation, and ideological leanings. After putting in balance the risks of false positives and the fact that the accounts had disappeared from the platform shortly after the vote, we decided to disclose the Twitter handles in the interest of accountability. We did so whenever there was a reasonable level of confidence that we were dealing with Twitterbots, to which ethical considerations of privacy are inconsequential. These concerns were also superseded by the public interest whenever we identified problematic content posted by bots. In other words, the tweets quoted directly for this project were expectedly posted by bots and are, therefore, of public and scholarly interest. In such contexts of large botnets participating in politically contentious debates, the ethical considerations regarding users' rights to not have their deleted tweet made public were deemed to be manifestly unfounded.

Campaign advocacy

For each tweet, we count the number of vocal hashtags advocating the Leave and Remain campaigns. The model thus relies on highly charged hashtags as a proxy for users' ideological affiliation. We tag the message as Remainer or Leaver based on the highest number of vocal hashtags used in association with each side of the campaign. Messages without hashtags advocating either side are tagged as neutral. The frequency count was aggregated to calculate the affiliation of users that tweeted or retweeted hashtags advocating either side of the campaign. Highly polarized messages – that is, tweets including several supporting hashtags – are, however, uncommon. For users championing the Vote Leave campaign, only 16 per cent of their messages included more than one such hashtag. These messages are yet more uncommon in the vote Remain campaign, where only 2 per cent of messages included more than one hashtag clearly associated with that side of the campaign.

We also identified the campaign affiliation (Leave or Remain) of users @-mentioned or retweeted in the original tweet. To achieve this, we loop through the data set to find messages tweeted by these recipients that championed either side of the campaign. We calculate the mode or 'mean partisan affiliation' per user based on the frequency of one-sided hashtags used throughout the period. The mean affiliation per user can only be calculated for users that actively participated in the referendum campaign on Twitter. In other words, for users at the receiving end (@-mentioned or retweeted) to be identified as Leaver or Remainer, the user in question must have tweeted or retweeted a separate tweet with hashtags clearly aligned with one side of the campaign, whereas a hypothetical user that may have tweeted an equal number of vocal Remain and Leave hashtags would be tagged as neutral. The rationale for restricting the parameters of ideological identification between users was to avoid mainstream media and high-profile accounts, which are regularly @-mentioned or whose tweets are retweeted with the addition of one-sided hashtags, to be classified in either side of the campaign battle. The mean affiliation has the added benefit of filtering out retweets or @-mentions intended as provocation or ironic remark; these messages are offset by the broader ideological orientation tweeted by the account, and users that have only sourced information or received @-mentions are classified as neutral for not having themselves tweeted any partisan hashtag.

This is a conservative approach to identifying campaign affiliation at the receiving end of a tweet, as users are only associated with one side of the campaign if the users themselves tweeted a partisan message at some point during the campaign. We believe this approach grounded on the mean affiliation per user reflects strong campaign membership with low probability of false positives. These conservative parameters to identifying campaign affiliation further reduced our data set to 33,889 tweets, the unit of analysis used in this study, posted by 15,299 unique users. Ultimately, the multiple sampling of the data (timespan, geographic location, campaign affiliation of sender and receiver) rendered a highly curated data set comprising ideological markers and geographically enriched data. While these procedures reduced the universe of tweets in the database dramatically, we believe this data set offers a defensible if limited representation of the debate, and our conclusions are conditional on these constraints.

We also probed whether ideologically motivated messages were more likely to disappear by classifying tweets that espoused nationalist versus cosmopolitan values, and populist versus policy-oriented values. This ideological value space leverages Inglehart and Norris's (2016) thesis of economic insecurity versus cultural backlash, which arguably accounts for the political realignment in Western political parties. To this end, we relied on a set of 10,000 tweets manually classified along the ideological polarities of Globalism vs Nationalism and Economism vs Populism to train a machine learning algorithm using text vectorization (Selivanov, 2016), an approach purposefully built for text analysis. For each ideological pair, the classifier returns a range of values from 0 (completely globalist and/or economist) to 1 (completely nationalist and/or populist), so that values from 0.45 to 0.55 are somewhere in the middle of this scale and assumed to be relatively neutral (Bastos and Mercea, 2018a). We also rely on a supplementary database – that is, a non-Brexit data set – to compare tweet decay in the Brexit data with non-political, generic hashtags that also trended in 2016. The non-Brexit data set includes 45 hashtags (see Figure 9.2 for details) and a random sample of 1,000 tweets was rehydrated to verify if user accounts and tweets were still available in the platform.

We followed the directionality of the information to graph a network of @-mentions and retweets, with $A \to B$ when B retweets A and $A \to B$ when A mentions B. We operationalize echo chambers as a function of the identified campaign affiliation. We tag each tweet as in-bubble if sender and receiver (@-mentioned or retweeted) have tweeted the same campaign. We tag the tweet as cross-bubble if the sender has tweeted one campaign and the receiver (@-mentioned or retweeted) has tweeted the opposite campaign. We tag the tweet as out-bubble if either sender or receiver was classified as neutral, which means any of them have not tweeted messages with clearly vocal campaign hashtags. Last, we deployed a bot detection protocol (Bastos and Mercea, 2018b) that led to the identification of 13,493 users with suspicious bot activity. To control for potential issues associated with bot activity, we replicated the analysis without this group of users, but the test did not yield significantly different results.

Geographic location of users

To leverage the granularity of our data, we rely on previous research that successfully mapped the UK EU membership referendum – restricted to local authority level – to parliamentary constituency level using a scaled Poisson regression model that incorporates demographic information from lower level geographies. This approach relies on a principled method of areal interpolation to aggregate the results at ward or constituency level, along with voting estimates at the level of council wards for authorities that have not disclosed the results at such granular levels (Huyen Do et al, 2015; Hanretty, 2017). The processed referendum data are thus relatively granular, with data down to the ward level in England, Scotland, and Wales. As the ward system does not exist in Northern Ireland, the data were aggregated at the local authority district, thus overcoming inconsistencies between local authorities and successfully mapping postcodes to parliamentary constituencies. We therefore adopt ward-level data when available, and estimates of the referendum results where ward-level data were not made available by the authorities. Such estimates advanced by previous research (Hanretty, 2017) allow us to investigate the extent to which the geographic distribution of tweets supporting each side of the campaign interacted with voting patterns across constituencies observed in the referendum.

Mapping geographically rich social media data onto census area or electoral districts is challenging due to the hierarchical subdivision of UK local government areas into various sub-authority areas and lower levels such as enumeration districts. As council wards comprise the most granular level to which we could retrieve results or estimates for the referendum vote, we sought to map referendum-related Twitter activity to this unit of geographic analysis. Therefore, we geocode and reverse-geocode the location of users that tweeted the referendum, and subsequently match postcodes to wards and parliamentary constituencies using the database provided by National Statistics Postcode Lookup (ONS Geography, 2011, 2017). Twitter users are thus simultaneously matched to the fields OSLAUA, OSWARD, and the PCON11CD (local authority, ward, and constituency codes, respectively). The first field includes local authority district (LAD), unitary authority (UA), metropolitan district (MD), London borough (LB), council area (CA), and district council area (DCA). Where the council ward system does not exist (that is, Northern Ireland), data were aggregated using these authorities to cover the entirety of the United Kingdom.

Upon geocoding the self-reported location of users, we found that only 30 per cent of them were based in the UK, with 19 per cent of users that participated in the Brexit debate based in the US and nearly 30 per cent in other EU countries. Also surprising is the large geographic spread of the British Twitter userbase, with London accounting for 14 per cent, Lancashire 7 per cent, Kent, Essex, West Yorkshire, and West Midlands ranging 3–4 per cent, and South Yorkshire, Hertfordshire, Cheshire, Merseyside, Surrey, and Hampshire at 2 per cent each. Taken together, each of these geographic groups is of comparable size to London in the share of users that tweeted the referendum.

We ultimately consolidate referendum and Twitter data based on OSLAUA (Local Authorities) and PCON11CD, which is the standardized ID code for each parliamentary constituency, the only GSS (Government Statistical Service) beyond European electoral region that is available for Northern Ireland and is consistent across the four countries included in the United Kingdom (ONS Geography, 2017). Using postcode as the common geographic marker across databases, this last step of data aggregation allows for pairing Twitter and referendum data based on LAD, each comprising a range of postcodes. We assigned pseudocodes when no postcodes or grid reference were made available by the authorities, particularly in the cases of the Channel Islands and Isle of Man. Data provided by the

Office of National Statistics assign the range E06 (UA), E07 (LAD), E08 (MD), and E09 (LB) to England; W06 (UA) to Wales; S12 (CA) to Scotland, and N09 (DCA) to Northern Ireland, with the pseudocodes L99 being assigned to the Channel Islands and M99 to the Isle of Man. Following these procedures, we first calculate the user-average score returned by the Brexit Classifier (Globalism, Economism, Nationalism, and Populism) and the mean campaign affiliation based on vocal hashtags tweeted by users.

It must be noted, finally, that identifying the location of social media users is a notoriously challenging task given the multitude of geographic information made available by social media platforms with various levels of accuracy, reliability, and granularity. While only 1 per cent of tweets usually include geolocational information (Sloan et al, 2013; Sloan, 2017), we have maximized this source by relying on the Twitter REST API to collect the 3,200 messages available per user, and searched for geolocation information in their tweets. Upon identifying a positive match, we apply this location to tweets authored by the same user that lacked geographic information. This approach maximized precision in determining the location of users, but there is no way of knowing whether the geolocation refers to a place where the user works, studies, lives, or was simply traversing. The same ambiguity pervades information made available in the user profiles, which furthermore may be entirely fabricated. In view of these caveats, we do not expect the geocode and profile data to necessarily reflect users' home or work location. Instead, we rely on this signal as geographic markers between two users sharing political information (homogeneous or otherwise). These procedures nonetheless allow for exploring the interaction between different geographic locations and ideological affiliations, as opposed to surveying the residence of Twitter users in the United Kingdom.

PART I

The Great Upset

THREE

Political Realignment

This chapter takes stock of the socioeconomic and cultural realignment in British politics underpinning the 2016 UK EU membership referendum, including economic inequality and a cultural backlash by older, traditional, and less educated voters. This cultural tension maximized political cleavages and deepened the wedge separating culturally divisive issues that would drive a realignment in the British political value space where the Conservative Party embraced nationalist rhetoric in response to European integration and immigration, whereas the Labour Party consolidated its grip on the liberal, educated, and affluent metropolitan elite clustered in and around London. We take stock of these two hypotheses and discuss the extent to which this political backdrop is reflected on our large database of Brexit tweets.

Cognitive dimensions

The referendum held in Britain to determine whether the country should remain a part of the EU marked the culmination of decades of political debate regarding the UK's membership in the supranational organization (Becker et al, 2016). The British electorate was marginally in favour of leaving the EU and the outcome of the vote opened a new chapter in the political life of the country (Asthana et al, 2016), which then embarked on a long and troubled process of redefining its relationship with the EU. The referendum also highlighted emerging ideological fault lines among the British electorate, broadly divided along economic and cultural dimensions, with the former dimension dominating from the 1970s to the early 1990s and the latter becoming increasingly prominent in the late 1990s and early 2000s (Kriesi and Frey, 2008).

A qualified discussion of the Brexit debate requires taking stock of the socioeconomic and cultural realignment in British politics that led to the result of the referendum, including economic inequality and a cultural backlash by older, traditional, and less educated voters. This led to a significant divide in the political landscape of the country, with deepening cultural and ideological differences between the main parties. The Conservative Party adopted nationalist rhetoric in response to European integration and immigration, while the Labour Party consolidated its support among the liberal, educated, and affluent metropolitan elite, mainly clustered in and around London (Bastos and Mercea, 2018a). These political cleavages continue to divide the Conservative and Labour parties and reflect a major realignment in British politics.

Economic insecurity and cultural backlash

The Scottish vote for independence may have anticipated a potential shift towards civic nationalism, but it was the Brexit vote that was largely driven by nativist and populist sentiments, which were defined in opposition to evidence-based policies and international cooperation. These economic insecurity and cultural backlash hypotheses were presented by Inglehart and Norris (2016) as a driving force in the shift towards nationalist and populist politics, with the possibility that these forces may interact and mutually reinforce each other. While the cultural backlash hypothesis suggests that certain sectors of the population became increasingly inward-looking and resistant to progressive value change, the economic insecurity argues that the decline of the blue-collar working class and lack of access to public services such as health, education, housing, and social welfare can explain the political upheaval.

These hypotheses are, of course, not mutually exclusive. Indeed, Inglehart and Norris (2016) argued that socioeconomic hardship and resistance to cultural change can reinforce each other. This has led to a cleavage between the young and well-educated who embrace progressive post-materialist values, including gender, sexual, and racial equality, but also human rights, environmental protection, secularism, and a greater tolerance of migrants, and the older and less educated who experience a decline in their material conditions and perceive a gradual erosion of values associated with industrial societies and solidarity around socioeconomic positions, religion, race, and geographic location. This section of the UK population perceived the cultural politics of identity recognition as a threat to traditional values, with immigration and the ensuing cultural change quickly exacerbating their disaffection. The EU embodied this cultural threat posed by other European societies, which was felt most acutely among people with low levels of education and income, manual workers, and the unemployed (McLaren, 2002).

The outrage over material inequality may have been compounded by a cultural backlash that was ultimately used and intensified by populist parties and leaders (Inglehart and Norris, 2016). Demographically, those who voted to leave the EU tended to be 60 years of age or older and with no educational qualifications (Becker, et al, 2016). The socioeconomic variables modelled by Inglehart and Norris (2016) support the economic insecurity hypothesis, which suggests that the decline in the fortunes of the blue-collar working class, coupled with a lack of access to public services such as health, education, housing, or social welfare, led to an extensive sense of insecurity. This hardship was attributed to the inability of political leaders in the country to spread the economic benefits of an increasingly integrated and relatively successful global economy (Piketty, 2014).

Maximizing political cleavages

Scholarship suggests that the British population considers social media as a critical component to their vote, but they continue to be ranked lower compared to other news sources covering elections (Dutton et al, 2017). The sense that social media have nonetheless reshaped the media landscape, with tangible consequences for democratic politics, stems from the argument that, either through intentional or incidental exposure via algorithmic filtering, users are narrowly exposed to information that reinforces their political beliefs (Sunstein, 2007). This selective exposure can entrench ideological polarization and hinder rational deliberation (Dahlgren, 2009). While some studies have shown that social media do provide exposure to diverse political views (Bakshy et al, 2015; Fletcher and Nielsen, 2017), the social dissemination of political content remains more likely among ideologically similar sources (Barberá et al, 2015).

This scholarship highlights the impact of social media on the polarization of audiences, which may in turn compound divisive issues across the electorate. Indeed, political polarization within parties based on cultural issues and social identities may have augmented the economic insecurity and cultural backlash experienced by sectors of the UK population (Inglehart and Norris, 2016), a development that maximizes political cleavages and deepens the wedge separating culturally divisive issues. The Conservative Party, in particular, has embraced cultural cleavages by adopting nationalist rhetoric in response to European integration, immigration, devolution, rising secularism, and declining influence in world politics (Kriesi and Frey, 2008).

Kriesi and Frey (2008) also argued that Labour and Conservative supporters converged on a liberal outlook on the economy up to the 1990s, a point after which Labour voters became more culturally liberal while Conservatives adopted more traditional conservative values. During this time, Labour consolidated its foothold among the highly educated and the middle classes, whereas the Conservative Party attracted the least educated and the working classes through a combination of nationalism and cultural conservativism (Kriesi and Frey, 2008), a process further exacerbated by the rise of populist parties such as the UK Independence Party (UKIP), which embraced traditional values and xenophobic appeals, and rejected outsiders and modern gender role leaders (Inglehart and Norris, 2016).

It was in this context of a changing political landscape that the use of social media and data analytics was portrayed as having aided effective campaigning, including the more accurate canvassing of potential voters, by groups such as the anti-establishment Vote Leave campaign (Cummings, 2016). Similarly, Celli, Stepanov, Poesio, and Riccardi (2016) leveraged social media data to predict the outcome of the 23 June vote based on users' expressed agreement with either the Leave or Remain campaigns on Twitter.

The Brexit value space

This research has led to the creation of a conceptual model that opposes globalism to nationalism and economism to populism. There are three notable definitions of populism that have guided social science research (Mudde and Rovira Kaltwasser, 2012). The first is a movement that is inspired by a charismatic leader and that cuts across class boundaries. The second is a way of doing politics that prioritizes the relationship between political leaders and the electorate over political parties. The third is a political discourse that claims to represent the experiences and beliefs of an oppressed majority in opposition to a hegemonic minority (Laclau, 2005). These definitions have been challenged as political parties started to adopt catch-all approaches to maintain their relationship with voters (Kriesi and Frey, 2008), a process that conflated populism with demagoguery but also took it to the centre-stage of party politics.

Despite these developments that limited the analytical value of the concept to the point of being employed as a floating signifier, the three definitions of populism pivot on the variance between 'the people' and 'the establishment' (Mudde and Rovira Kaltwasser, 2012). Populist discourse often highlights the socioeconomic, political, and cultural marginalization experienced by certain groups. This discourse portrays the antagonism as a physical distance between the centres of political and economic power, such as Brussels, London, or the South-East of England, and other regions of the UK (Wills, 2015). This discourse characterizes those in power, often corporate or governmental, as privileged and disconnected from the realities of ordinary people. This elite is accused of promoting economic and political globalization that is represented by the EU and European integration (Kriesi et al, 2008; Woods, 2009).

Nationalist parties took advantage of the dissatisfaction caused by economic trade agreements, technological changes, and the idea of a culturally and ethnically diverse nation. This political value space projected a state that is homogeneous in terms of culture and ethnicity, and that would ultimately regain its power by opposing trade liberalization agreements and controlling labour migration flows that put pressure on the welfare state (Mudde, 2000, 2004). On the other hand, globalism, the third dimension of the latent ideological space, is characterized by a universalistic worldview that emphasizes individual rights, and views citizens as free agents operating in a global economy where national political systems are increasingly similar to one another (Turner, 2002).

In its more basic forms, populist messages advance a discontent directed at elites and the establishment and foreground popular will, while nationalist sentiments revolve around notions of national exceptionalism, sovereignty, and nativism (Parker, 2016). In opposition to that, a prevalent response to popular disenchantment has been a drive towards greater efficiency in economic policy and analysis, policy making, and government administration. This is the fourth coordinate in the latent ideological space which we term economism, conceived in opposition to populism and that emphasizes consensus building, due process of law, and accountability, but also expert analysis and evidence-based policy making that could drive consensus across ideological fault-lines (Nilsson and Carlsson, 2014). As such, economism refers to the comprehensive political consensus to safeguard free market economics embodied in government policy and the array of expert bodies – from the Bank of England to think-tanks, business, and trade organizations – which have helped define and uphold it in the last three decades (Crouch, 1997). Figure 3.1 unpacks these ideological coordinates of the Brexit political value space.

Figure 3.1: Ideological coordinates of British public opinion and political value space

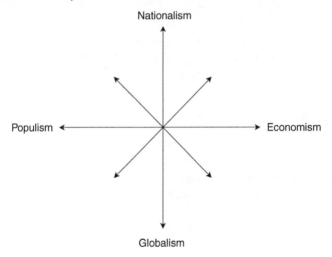

FOUR

Nationalism and Populism

In this chapter we discuss nationalism and populism in British politics and review the literature on the cultural cleavage that incorporated nationalist rhetoric in response to European integration and immigration. We test Inglehart and Norris's dual hypotheses of economic insecurity versus cultural backlash as key developments underpinning the upsurge in nationalistic and populist sentiments, the former foregrounding the economic decline of the blue-collar working class, and the latter arguing that sectors of the population have become increasingly inward-looking and averse to progressive value change. This chapter also unpacks the political value space used to train a machine learning algorithm and the resulting analysis of the Brexit tweets.

The geography of Brexit

The Brexit vote was characterized by significant regional differences in voting patterns. The Leave vote was strongest in economically deprived regions outside of London, including the North-East and the English Midlands, but also in Wales. In contrast, London and other affluent metropolitan areas voted strongly in favour of Remain. There were also significant differences in voting patterns between an increasingly independent-minded Scotland and Northern Ireland, both of which voted to remain in the EU, and England and Wales, which voted to leave. The situation in Northern Ireland was further complicated by the potential implications of Brexit for the border with the Republic of Ireland, which remains a member of the EU, a point of contention that would remain unresolved long after the vote (Rennie Short, 2016).

The prevailing sociological explanation for the uptake of populist and nationalistic politics driving Brexit and the election of Donald J. Trump in the US were encapsulated in the dual hypotheses presented by Inglehart and Norris (2016). This contrasted the economic insecurity with the cultural backlash hypothesis as key developments underpinning the upsurge in nationalistic and populist sentiments. The economy insecurity hypothesis foregrounded the economic decline of the blue-collar working class, whereas the cultural backlash hypothesis argued that sectors of the population have become increasingly inward-looking and averse to progressive value change. The split was also defined culturally, with liberal political values supporting environmental protection and European integration promoted by the Labour Party that contrasted with traditional values embodied in the pursuit of 'law and order' and a concern with immigration heralded by the Conservative Party (Kriesi and Frey, 2008).

These socioeconomic factors brought about a political realignment that had been happening for some time, but which was not noticeable due to the UK's majoritarian electoral system that maintained the two main political parties in power (Dunleavy and Margetts, 2001). Historically, the British electorate was traditionally divided along economic (dominant from 1970s to 1990s) and cultural dimensions, which became increasingly prominent in the late 1990s and early 2000s. On the economy, voters differed in their support for the welfare state or its liberalization, with a left–right compromise that included some protections for the welfare state in exchange for economic liberalization (Kriesi and Frey, 2008).

Nationalist populist parties tapped into a widespread feeling of disenchantment and scepticism towards the elite, combined with a desire for cultural and ethnic values that emphasized the role of the state in opposition to trade liberalization and loss of control over migration flows (Mudde, 2000, 2004). The targets were the privileged and rootless cosmopolitans pandering to economic and political globalization personified by the EU and European integration (Kriesi et al, 2008; Woods, 2009). Therefore, the British decision to leave the EU reflected a cultural backlash by older, more traditional, socioeconomically deprived, and less educated voters or alternatively reflected the broad sense of economic insecurity experienced by blue-collar working-class voters (Becker et al, 2016; Inglehart and Norris, 2016). These explanations were often seen as compounding factors, rather than mutually exclusive hypotheses, in the manifold sociological explanations of Brexit.

Brexit parameters

The Brexit tweets leave little doubt that nationalism and populism sentiments, whether driven by cultural cleavage or economic insecurity, were integral to the Brexit agenda rhetoric in response to European integration and immigration. In our study 'Parametrizing Brexit: Mapping Twitter Political Space to Parliamentary Constituencies', that appeared in *Information, Communication & Society*, we introduced a political value space used to train a machine learning algorithm that could empirically test these hypotheses with the Brexit tweets database (Bastos and Mercea, 2018a). This was achieved by devising a machine learning algorithm based on a score assigned to each tweet across two scales: the first with nationalism at one end and globalism at the other, and a second with populism and economism at either end.

The data used to train the deep learning algorithm were restricted to the period of the referendum campaign, namely between 15 April 2016 and 24 June 2016. The posts were authored by 30,122 unique UK users, whose geographic locations were traced up to postcode level. All 72 constituencies with overwhelming support for Vote Leave (65 per cent voting to leave, or higher) presented predominantly nationalist sentiments. Conversely, only 17 of these constituencies had a Twitter debate predominantly defined by populist sentiments, with the other 55 being classified as economist. These are regions in which Brexit was wholeheartedly embraced and yet populist sentiments were not found to be predominant in the Twitter data stemming from these regions.

This approach allowed us to look at how often these sentiments were expressed during the campaign and how attitudes expressed on Twitter for each constituency matched the result of the vote. The results were interesting, if surprising. After mapping more than half a million tweets across the UK to 650 parliamentary constituencies, we found that opinions posted on Twitter accounted for almost half of the voting patterns seen in the actual referendum once demographic data for each region were taken into account, such as unemployment, valid votes, size of electorate, and ratio of retired population living in the constituency, which are variables known to be associated with Leave and Remain voters. We also found that the Brexit discussion on Twitter, at least for the cohort of users based in the UK, was driven by economic and nationalist politics, and less so by populist and globalist issues.

Three-quarters of the messages (74 per cent) displayed nationalist sentiments, such as the desire for the country to be self-governing, as opposed to 26 per cent that expressed globalist values, such as universal individual rights and international cooperation. Almost two-thirds of tweets (62 per cent) focused on economic issues underpinning Brexit, such as trade policy, instead of expressing populist sentiments (38 per cent), such as voters taking back control from elites. The machine learning algorithm identified a prevailing nationalist sentiment, which persisted throughout the campaign and was only offset in the last days when a globalist upsurge neatly divided the British Twittersphere into nationalist and globalist sentiments.

What was particularly surprising was that tweets embracing nationalist content did not originate from economically fragile areas that were generally supportive of Brexit, such as northern England, but from various other regions across the country, including remain-backing areas such as Scotland. The results were somewhat at odds with previous studies that highlighted the importance of populist ideas and global issues in the outcome of the referendum (Celli et al, 2016). While nearly 40 per cent of tweets presented populist sentiments, these messages were concentrated in a small number of constituencies. Only 10 per cent of the parliamentary constituencies presented prevailing populist sentiments, compared with economic issues, and less than 5 per cent presented globalist, compared with nationalist sentiments.

As such, our first overture through the Brexit database entailed modelling the data for populist, nationalist, globalist, and economist sentiments. We explored whether political talk on the British Twitter Monthly Active Userbase could quantifiably mirror this process. Specifically, we examined the relationship between tweets and the electoral geography of the Brexit referendum to assess the extent to which users tweeting nationalist and populist content would overlap across geographic enclaves; and conversely, whether such a pattern could be observed in relation to users tweeting globalist or economist content. In other words, we probed whether the Twitter public stream can be used to identify, measure, and model the political consequences of an alignment between the vote and broader ideological orientations expressed by the British public opinion. This study was important because it yielded a machine learning algorithm (the 'Brexit Classifier') that would ultimately be applied to the entire database of Brexit tweets.

Brexit Classifier

The Brexit Classifier is a machine learning algorithm that resulted from multiple tests to identify the four key ideological coordinates theorized by Inglehart and Norris (2016). We relied on two expert coders who classified 10,000 tweets along the ideological coordinates of Globalism, Economism, Nationalism, and Populism. We controlled for intercoder reliability by double-coding a random sample of tweets (N=100) repeatedly, and after four rounds we achieved a Krippendorff's alpha of 0.94 for the complete value space, with alpha of 1 for the Globalism-Nationalism dyad and 0.86 for the Economism-Populism polarity. We relied on this trained set of tweets to parametrize the machine learning algorithm using text vectorization (Selivanov, 2016), an approach that converts text to a vector representation and generates embedding vectors through deep learning.

Unlike frequency-based approaches to text classification, which simply compute the number of positive and negative words (or hashtags) and draw a conclusion based on the final sum, text vectorization is a deep learning algorithm that draws context from phrases. It is often deployed to analyse and classify large text corpora, including user feedback, reviews, and comments. The deep learning algorithm can handle linguistic variation and performs well with misspelled or poorly constructed sentences, a marker of Twitter communication, because it considers the entire body text of tweets to infer ideological inclination. It is independent from hashtags, though in the Brexit corpus we found hashtagged tweets to be more vocal and likely to display a clear alignment with one of the four ideological coordinates. As a result, and unsurprisingly, the algorithm consistently identifies campaign hashtags as valid indicators of tweets ideologically leaning towards a given position in the political value space.

Training a machine learning algorithm is essentially a trade-off between recall, the number of correct results divided by the number of possible results, and precision, the ratio of positive and relevant matches. In other words, the more variables the algorithm has to identify (in our case there are four: globalism, economism, populism, and nationalism), the higher the likelihood that the algorithm will be unsuccessful. In our case, the algorithm needed to identify at least one and a maximum of two ideological coordinates, as the polarities globalism-nationalism and economism-populism are mutually exclusive. Given the disjoint assumption of the political value space, we maximized precision and recall by splitting the ideological value space along two polarities and training two separate algorithms later combined into a single classifier (henceforth, the Brexit Classifier). This approach successfully returned substantially more relevant results, while also returning most of the relevant results.

We relied on the aforementioned set of 10,000 manually coded tweets to assign a value (positive or negative) to each of the concepts discussed by Inglehart and Norris (2016), with the algorithm calculating the probability of positiveness and negativeness for each ideological polarity (Globalism vs Nationalism and Economism vs Populism). For each ideological pair, the classifier returns a range of values from 0 (completely globalist) to 1 (completely nationalist), so that values from 0.45 to 0.55 are somewhere in the middle of this scale and assumed to be relatively neutral. The algorithm was trained using Document-Term Matrix (DTM), vocabulary-based vectorization, and the TF-IDF method for text pre-processing. Figure 4.1 shows the area under the curve (AUC) on train and test data sets for the Economism-Populism and Globalism-Nationalism ideological pairs (AUC=0.8697 and AUC=0.901, respectively). The algorithm performed well for the set of 565,028 tweets posted during the campaign period and that included geographic information. In the last step of the classification, the algorithm calculates the best fit, projects the results along spatial coordinates comprising the four ideological dimensions, and estimates significant oscillations between any of the ideological pairs.

The algorithm, in summary, allowed for calculating the mean campaign affiliation, mean globalist-nationalist, and mean economism-populism for each user that tweeted the referendum. By matching users to local authority districts, we managed to identify the relative distribution of nationalist and populist sentiments across the UK. Twitter data were, therefore, aggregated first at user level, and subsequently at

Figure 4.1: Area under the curve calculated from train and test data sets for the Economism-Populism and Globalism-Nationalism ideological pairs

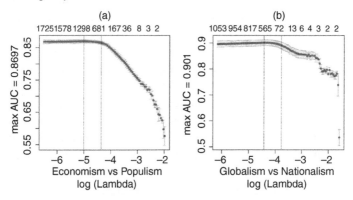

constituency level, which was the unit of analysis employed. The resulting data included multiple streams of Twitter data consolidated into a single database of online and offline activity at the constituency level: first, their ideological alignment on Twitter, and second, their voting preferences relative to the 2016 UK EU membership referendum.

It must be noted that there are important limitations to this approach. First, during the process of training the classifier we struggled to separate economism from populism, as many of the populist claims are economic in nature. This is reflected in the lower AUC score for Economism-Populism compared with Globalism-Nationalism. We addressed this challenge by accentuating the policy and expert-oriented component of the economism polarity, which sits in opposition to populist views that appeal to emotion and the perceived rights 'of the people', a value that is difficult to pinpoint but that stands visibly against the value space occupied by economism. Second, the clear identification of messages with nationalistic content has limited heuristic value, as nationalistic sentiments in Scotland and England refer to fundamentally different political agendas.

Globalism, nationalism, economism, and populism

Our analysis explored the political geography of the vote at the level of local authority areas, a level that was similarly explored in a range of other studies. Becker et al (2016) used an Ordinary Least Square (OLS) regression model with a best subset selection machine learning protocol to identify a set of factors associated with the outcome of the referendum. Although they suggested that a higher turnout in urban areas could have resulted in a different outcome, the authors noted that the vote to leave the EU was positively correlated with support for the Eurosceptic UK Independence Party (UKIP) and the British National Party in the 2014 European Parliament elections. Other factors that were strongly associated with a vote to leave included employment in the manufacturing sector, lower hourly wages or higher unemployment rates, a higher proportion of rented council housing in the area, longer waiting times for access to public health services, and lower levels of public sector employment.

We modelled the Brexit ideological value space by testing the Globalism-Nationalism (henceforth, GlobNat) and Economism-Populism (henceforth, EconPop) generated by the Brexit Classifier against the RemainLeave variable, which measured campaign support based on vocal hashtags, and ultimately against the result of the referendum. The distribution of Nationalistic and Globalist tweets mirrored the distribution of Remain and Leave votes across parliamentary constituencies in the UK, but it was the EconPop variable that proved particularly effective. While both GlobNat and EconPop are significantly correlated with the results of the referendum ($r=$ 0.31 and 0.46, respectively, $p<.0001$), the polarity Globalism-Nationalism had only a modest explanatory power ($R^2_{adj}=.10$, $p=9.226\text{e-}16$), whereas the polarity Economism-Populism could explain over one-fifth of the variance found in the results of the referendum ($R^2_{adj}=.21$, $p<2.2\text{e-}16$). Figure 4.2 shows how the classifier positioned each of the half a million

Figure 4.2: Ideological value space calculated from Twitter messages

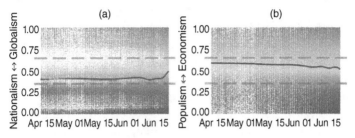

Note: The solid lines indicate the probability of Nationalist versus Globalist (a) and Populist versus Economist (b) sentiments, respectively. Plotted dots indicate the position of each of the half a million messages.

tweets processed in this study with the fitted line representing the trend detected by the algorithm.

The classifier also found a strong nationalist sentiment that persisted throughout the campaign and was only offset in the last days when a globalist upsurge brought the British Twittersphere closer to an equally partitioned divide between nationalism and globalism. For most of the campaign, the overall sentiment was decidedly nationalistic, averaging .40, which translates to three-quarters of messages having a nationalistic sentiment. On the Economism vs Populism spectrum, the sentiment was reversed: most messages tweeted in the period (61 per cent) were dedicated to economic implications of the decision to leave the EU. Though messages with a strong populist appeal accounted for less than 40 per cent of the total messages, the trend shown in Figure 4.1 is of growing occurrence of populist messages in the weeks and days leading up to the vote, with messages centred on economic issues moving out of the debate as a populist discussion balloons.

When combining demographic data with the political value space mapped with Twitter messages, the model managed to account for nearly half of the variance found in the referendum results ($R^2_{adj}=.41$, $p<2.2e-16$). While social media data remain a non-representative sample of the larger population, they can provide important markers for understanding the evolution of public debates and the geographic coverage of the discussion (Bastos et al, 2014). The results of the classifier also shed light on the importance of economic issues that may have motivated the userbase tweeting the referendum, a component of the Brexit debate overshadowed by the much-discussed cleavage between the metropolitan elite in London and parts of England and Wales that were economically worse off. Figure 4.3 shows the scores for GlobNat, EconPop, and the results of the referendum across parliamentary constituencies.

However, more than half of the variance in the referendum results remained unaccounted by the model, and a closer inspection of the aggregate scores for Globalism, Nationalism, Economism, and Populism shows that the map only partially matches the results of the referendum (Figure 4.3c). Apart from London and north-west Wales (Gwynedd), globalist messages are absent in Figure 4.3a, with nationalist content appearing in Scotland (which voted Remain but has long contended with

Figure 4.3: Mean score of (a) Globalism-Nationalism and (b) Economism-Populism for each parliamentary constituency, and (c) the results of the referendum

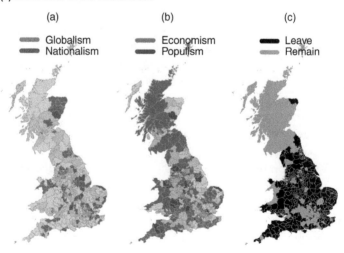

a nationalistic agenda pressing for an independent Scotland), the English Midlands, and the North of England. Populist messages are also relatively underwhelming, covering only portions of the Midlands and the North (Figure 4.3b). It is the economic discourse that is prevalent in the debate registered in the Twittersphere, being particularly prominent in Scotland, north-west Wales, and Greater London (Figure 4.3b). As an expression of the public opinion, Twitter debate appeared invariably focused on economic and nationalistic issues as opposed to the populist and globalist sentiments thought to have shaped much of the referendum campaign (Inglehart and Norris, 2017).

As such, we did not find, for one, that economically fragile northern Britain was any more likely to embrace nationalist content. In fact, it was Scotland that appeared as a relatively fertile ground for nationalist messages. The distribution of globalist, nationalist, populist, and economist content was also found to be somewhat at odds with the geographic distribution of the Leave-Remain vote. The conversation on Twitter in the weeks leading up to the referendum vote was largely centred on nationalistic and economic sentiments, indicating that not only material inequality, but also ideological realignments have contributed to the outcome of the referendum. On the one hand, the variables that improved the model are associated with issues surrounding material inequality, chief of which are the percentage of economically active residents and the size of the parliamentary constituency. On the other hand, ideological orientation also proved capable of explaining the unexpected outcome of the UK public's vote to leave the EU.

There was plenty of evidence in the data indicating that nationalism was a quintessential component of the referendum debate during most of the campaign, with three-quarters of messages having some degree of nationalistic sentiment embedded in them. These results, however, need to be considered in the context of the limited heuristic value of this ideological coordinate. While the classifier successfully identifies nationalistic sentiments in Scotland and the Midlands, these areas have voiced fundamentally different versions of nationalism. In other words, although nationalism appears to have been a critical marker of the Brexit value space, there are major differences in Scottish and English nationalism that extrapolate the heuristic confines of the classifier. Populist messages, however, were decidedly of lesser importance compared with the sheer volume of tweets discussing the economic consequences of Britain leaving the EU, a trend that was, however, overturned as we approached the date of the vote.

PART II

Bots Talking to Bots

FIVE

Polarization

In this chapter we take stock of the context of polarized politics and hybrid media environments undergirding the Brexit vote, where news stories were discussed, disseminated, and interpreted within politically homogeneous online communities secluded from diverse ideological information. We discuss echo-chamber communication in the Brexit debate against the backdrop of divisive and affective polarization, with significant differences in the Leave and Remain campaigns that speak to the fundamentally different social networks to which these groups are attached. The chapter also unpacks the often-confused concepts of echo-chamber communication and filter bubbles and reports on methods to identify politically homogeneous communication online and offline.

Polarization

The idea that online activity has no impact on real-world events is based on the belief that virtual communities are not connected to offline social networks. This assumption claims that online activism is ineffective and does not result in any concrete actions, as individuals express their discontent solely within online circles, which are incapable of enacting change in the real world. This argument forms the basis of the discussions on clicktivism and slacktivism, an assertion passionately advocated by Morozov (2013) and convincingly argued by Gladwell (2010), though a wealth of studies has since identified significant relationships between online activism and developments on the ground (Bastos et al, 2015; De Choudhury et al, 2016; Freelon et al, 2018). While these studies found significant associations between online and offline activity, they rarely included granular spatial data accounting for its geographic dispersion.

Social media platforms with global coverage run counter to traditional, offline communities where group dynamics follow clear boundaries defined by space and size typical of classrooms, colleges, workplaces, sports teams, churches, trade unions, and political parties. Online social networks are subject to no such spatial and social constraints typical of such arrangements, and have quickly outpaced the scale of mass media with communities that grow beyond this inflexible limitation. The absence of clear group boundaries necessary for in-group favouritism and outgroup derogation may trigger new forms of coalition and affiliation not prescribed in classic social identity theory (Tajfel, 1974); the response amplification may likewise lead to disruption and disintegration of traditional political and partisan allegiances along with divisive politics known as affective polarization (Iyengar et al, 2012, 2019).

The spatial embedding of social media interactions can foreground online social relationships notwithstanding the centrality of offline social networks that surround us. Despite this epochal change, studies continue to find that social network links are dependent on geography, with the probability of ties reducing with distance (Liben-Nowell et al, 2005; Wong et al, 2006; Mok et al, 2010; Preciado et al, 2012) and proximate actors having similar sociocultural and demographic properties (Hipp et al, 2014). There is also evidence that homophily offline may trigger similarly clustered formations online (Onnela et al, 2011). Indeed, a large body of scholarship has sought to model the probability of tie formation online as a function of geographic propinquity – that is, the hypothesis that people located closer together in physical space are more likely to form a relationship (Festinger et al, 1950).

There is also extensive work probing space-independent communities in spatial networks (Expert et al, 2011). In the seminal study of Onnela et al (2011), a social network of individuals with precise geographical information for each actor was collected. The variation of geographical span for social groups of varying sizes was surprising, as no correlation between the topological positions and geographic positions of individuals within network communities could be found. The results were at odds with scholarship positing a linear association between ties forged online and geographic location, including the established dependency of geography on the structure of dyadic social interactions, as friendship probability has been shown to decay with distance (Liben-Nowell et al, 2005). In contrast to that, the results reported by Onnela et al (2011) suggested that spreading processes may face distinct structural and spatial constraints.

The relative autonomy of online social networks can insulate individuals from nearby communities. Well-educated and otherwise open communities may conceivably grow apart due to asymmetric information diets feeding different in-group identities. Conflicting information diets across communities can feed motivated reasoning and confirmation biases that jeopardize consensus-driven communication, with climate change being a critical development where the abundance of information, or the scientific consensus about the subject, is not relevant to several communities (Latour, 2004; Bardon, 2019). Evidence points to ideological polarization actually increasing with respondents' knowledge of the subject, so that the likelihood of conservative people being climate change deniers is higher if they are college-educated (Bolsen et al, 2015). But disagreeing with facts advanced by the scientific community is not a prerogative of conservatives. While conservative people are more likely to reject established facts about evolution, the age of the Earth, and climate change, liberals are particularly prone to reject scientific facts about fracking, vaccination, and GMO. Liberals are also less likely to accept expert consensus on the possibility of safe storage of nuclear waste or on the effects of concealed-carry gun laws (Kahan et al, 2011).

Naturally, social groups formed online or offline are not static. There is, however, at least one key affordance of social platforms rendering online communication substantively different from face-to-face interaction. This is the potential for politically and culturally homogeneous communication. This is encapsulated in the metaphors of filter bubbles and echo chambers, which one may find difficult to route around, or even to escape in an information ecosystem marked by rampant social media use. This is a considerable departure from the social web marked by open standards and centred on user governance that dominated the internet studies in the nineties and early noughties. Where web portals would bring people of different affiliations together (Tyler et al, 2019), social platforms can fragment and Balkanize lifestyles and political allegiances, which further segregates and polarizes online communities. This process of increasing segregation and polarization online may then spill over to face-to-face communication through network externalities, including spillover effects (Bastos, 2021b).

Ideological clustering

The prevailing narrative about ideological clustering argues that the interaction patterns existing in social platforms lead users to engage with those who are politically aligned and share content that resonates with their ideological orientation (Sunstein, 2007). The ideological clustering observed in politically homogeneous echo chambers and algorithmic filter bubbles would stand in contrast to the diversity of opinions found in face-to-face interactions. These forces, unleashed by social media interaction and algorithmic filtering, would ultimately jeopardize political compromise, as ill-informed individuals move to a landscape marked by tribalism and information warfare (Benkler et al, 2018), a development enabled by a business model driven above all by the commodification of digital circulation and its capitalization on financial markets (Langley and Leyshon, 2017).

There is a substantive body of observational evidence showing the role of social media in stratifying users across information sources (Conover et al, 2011). While the rapid growth of online social networks fostered an expectation of higher exposure to a variety of news and politically diverse information (Messing and Westwood, 2014), they also increased the appetite for selective exposure in highly polarized social environments (Wojcieszak, 2010), with the sharing of controversial news items being particularly unlikely to take place in such contexts (Bright, 2016). The filter bubble hypothesis encapsulated these claims by positing that social platforms deploy algorithms designed to quantify and monetize social interaction, narrowly confining it to a bubble algorithmically populated with information closely matching observed and expressed user preferences (Pariser, 2012).

However, research had also challenged the notion that social media caused selective exposure or ideological polarization, the latter being reportedly more pronounced in face-to-face interactions (Horrigan et al, 2004; Gentzkow and Shapiro, 2011; Boxell et al, 2017). Exposure to diverse and even competing opinions on polarizing topics was found to occur on social media across various national contexts (Kim, 2011; Bakshy et al, 2015; Fletcher and Nielsen, 2017). Similarly, social media were shown to be coextensive with more diverse personal networks which are more likely to include individuals from a different political party (Hampton et al, 2011). Even with scanty evidence linking filter bubbles and echo chambers to general social media communication, there is compelling evidence of echo-chamber communication in several political contexts (Barberá et al, 2015).

One possible explanation for the conflicting evidence on echo chambers is that politically homogeneous communication might reflect group formations inherited from offline social relations. These homophilic preferences can coexist with social media platforms that provide ideologically diverse networks (Barberá, 2014). As such, the boundaries of one's network can be simultaneously permeated by echo chambers stemming from offline relationships while being exposed to competing opinions on polarizing topics that circulate on social media. Similar associational effects have been reported in the literature, with the relationship between spatial distance and users' interaction on social media found to be significant, and friendship ties in densely connected groups arising at shorter spatial distances compared with social ties between members of distinct groups (Laniado et al, 2017). More importantly, research found social ties on Twitter to be constrained by geographical distance, with an over-representation of ties confined to distances shorter than 100 km (Takhteyev et al, 2012). These geographic constraints are likely to interact with the geographic patterning of the Brexit vote, which reveals spatial and associational segregation, with a spatial distribution in which people are more likely to talk to those who are categorically more similar to them.

Literature on social media polarization has often lumped together dissimilar concepts such as echo chambers, filter bubbles, and information cocoons to refer to politically homogeneous interaction on social media and to ideologically homogeneous forms of news consumption facilitated by social platforms. This is unfortunate because echo chamber and filter bubbles refer to fundamentally different developments. The former is defined as a closed communication system, much like Facebook users who choose to interact mostly with users with whom they are ideologically aligned. Filter bubbles, on the other hand, at least in the definition advanced by Pariser (2012), describes a closed system resulting not from politically homogeneous interaction, but from personalized searches and website algorithms that increasingly select information congruent to one's political views. As such, echo chambers refer to closed communication systems where users interact with each other, as opposed to filter bubbles which refer to the algorithmic selection involved in the process of searching and retrieving information online.

The conceptual confusion is rooted in concerns about political polarization that encapsulate both echo-chamber communication and filter bubbles. The specialized literature often reports on findings restricted to incidental exposure but extrapolate the conclusions to polarization, thereby inadvertently conflating echo chambers with filter bubbles. The evidence for filter bubbles is nonetheless fraught. This is likely to be partially caused by the reliance on survey methods that are not suitable to investigate the problem. Survey methods, for one, cannot list all media sources individuals may be exposed to in the course of a day because there are, of course, too many to include, but also because the researcher may not be familiar with them all, some of which may in fact not yet exist at the time of the survey inception. As such, observation data gleaned from surveys are likely to ignore the long tail of disinformation when assessing news diets. These problems are compounded by the relative loose theoretical footing informing much observation research, with terms like echo chamber and filter bubbles often being used interchangeably. This is problematic because while echo chambers refer to politically homogeneous communication, filter bubbles refer to information exposure (incidental or otherwise).

Geographic propinquity

There is, however, a large body of scholarship reporting echo-chamber effects on social media (Wojcieszak and Mutz, 2009; Krasodomski-Jones, 2016; Vaccari et al, 2016). In these settings, political information was more likely to be retweeted if received from ideologically similar sources (Barberá et al, 2015), and cross-ideological information was unlikely to circulate in social clusters with a strong group identity (Himelboim et al, 2013a, 2013b), with clustering around partisan information sources being more prevalent among conservative than liberal voters (Barberá et al, 2015; Benkler et al, 2017). As such, there remain widespread concerns that the relative autonomy of online social networks can insulate individuals from nearby communities, and that otherwise open communities may conceivably grow apart due to asymmetric information sharing feeding different in-group identities.

Identifying echo chambers online is challenging because one needs to untangle politically homogeneous communications from the location where they emerge. This requires considering spatial non-adjacency in the interdependence of social relationships. One important consequence of the spatial constraints imposed on networks is the cost associated with the length of ties (edges), which in turn has substantive effects on the topological structure of networks. A result of this constraint is that for most real-world spatial networks, the probability of a tie between any two actors decreases with distance. Spatial constraints impinge not only on the structure of networks, but also on endogenous processes driving tie formation, such as phase transitions, random walks, synchronization, navigation, resilience, and disease spread (Barthelemy, 2014). This definition nevertheless ignores links that are not necessarily embedded in space. The very nature of the links in social relationships such as friendships, whether established online or offline, is a virtual network by definition. While the actors are embedded in space, the links connecting individuals are not planar and therefore are fundamentally dissimilar from perfectly two-dimensional spatial networks.

These problems are compounded by the difference between relationships established offline, where actors are embedded in geographic space, and online, where individuals may be entirely unaware of each other's location. The intricate relationships between physical ties and online interactions are captured in the problem of the directionality in social relationships, a sociological debate about the causal direction of homophily. Homogeneous social networks are marked by assortative social ties and result in limited social worlds that constrain their access to information, the attitudes they form, and the interactions they experience (McPherson et al, 2001). Similarity breeds proximity, which then reinforces homophilous tendencies within cultural, economic, ethnic, sexual, religious, and racial groups, which become more likely to interact or form social ties (Lazarsfeld and Merton, 1954). Causality may, of course, move in either direction, but researchers have focused mostly on the hypothesis that similarity causes interaction (McPherson and Smith-Lovin, 1987).

It is, therefore, important to separate echo-chamber from neighbourhood effects, often characterized with Miller's (1977) assertion that locality is a better predictor of how people vote than their social characteristics. Johnston and Pattie (2011) provided an account of how neighbourhood effects can trigger feedback loops and spillovers. For instance, if a political debate between individuals holding opposing views about the best candidate in an election can persuade people to reconsider their own positions (conversion), then any social network with a majority of the population supporting a given candidate is more likely to switch to the minority candidate than the other way around. As such, contact with voters in a given area influences not only individuals directly connected, but also others in the neighbourhood (Huckfeldt and Sprague, 1995). On the other hand, this process would render the majority view within the network even more prevalent, as dominance would be greater than expected from knowledge of the individuals' personal characteristics alone. If neighbourhood effects can be observed whenever conversation networks are spatially constrained, then the political complexion of areas should be more polarized than their social composition implies.

Similarly, social media microtargeting can spill over to actors in local social environments who were not targeted by the message themselves. In this scenario, the spillover effect is the contagion effect on actors due to interventions targeting their friends and acquaintances. In contrast, endogenous peer effects stem directly from peers. In other words, actors in a network influence each other without being subjected to any microtargeting or intervention. In the following, we explore the Brexit campaign on Twitter to unpack endogenous and exogenous peer effects, chief of which is geographic propinquity. These constraints interact with the geographic patterning of the Brexit vote, which reveals further spatial and associational segregation, with a spatial distribution in which people are more likely to talk to those who are ideologically more similar to them.

Brexit echo chambers

The Brexit tweets database allowed us to identify whether the ideological clustering of tweets was associated with physical, in-person interaction occurring in offline networks (neighbourhood effect) as opposed to being restricted to online networks (echo-chamber effect). The assumption that online echo chambers would be associated with geographic propinquity runs counter to much of the literature on echo chambers, where this process is defined as a self-selection that confines online communication to ideologically aligned cliques (Del Vicario et al, 2016; Zollo et al, 2017). Space is noticeably absent in this body of work addressing politically homogeneous echo chambers. Transposed to the network of tweets about the UK EU membership referendum, and following the prevailing narrative found in the literature, we would expect to find echo chambers as a communication artefact resulting from online discussion alone. Conversely, one would not expect the geographic locations of users to play a significant role in the formation of echo chambers, as echo chambers would result from social media interactions unfettered by geographic space.

This, however, is not what we found (Bastos et al, 2018). We identified significant geographic constraints driving politically homogeneous interaction that overlapped with the patterning of the Brexit vote. The results showed that although most interactions were within a 200-km radius, echo-chamber communication was largely restricted to neighbouring areas within a 50-km radius. Instead of resulting from interaction limited to social platforms, echo chambers seem to reproduce the structural political polarization found in offline social networks. The geography of echo chambers was, however, different between the Leave and Remain campaigns, with the former spanning much shorter distances compared with the latter. The trend was also reversed for non-echo-chamber communication, which covered shorter distances on the Remain side compared with echo-chamber communication. In other words, Leave campaign messages were chiefly exchanged within ideologically and geographically proximate echo chambers. While echo chambers also prevailed on the Remain side, the trend was however inverted: as distance between sender and receiver increased, echo chambers became more common and covered increasingly larger geographic areas compared to non-echo-chamber communication. Leave echo-chamber messages covered 199 km compared with 238 km for Remain. The average Leave message, similarly, covered 234 km compared with 204 km for Remain.

But the intensity of echo-chamber communication is remarkably similar on the Remain side of the campaign, where only 10 per cent of users directed @-mentions or retweeted content from users identified with the Leave campaign (cross-partisan or cross-bubble communication), with 22 per cent of interactions including neutral users (out-bubble communication), and a total of 68 per cent of interactions initiated by Remainers being echo chambers (in-bubble communication). The likelihood of users campaigning for one side of the referendum engaging with users of the same leaning – instead of neutral or adversarial users – was captured by fitting a linear regression model on the sender's affiliation as the explanatory variable of echo-chamber communication: partisan affiliation explains nearly half of the variance in the data (R^2_{adj}=.44, p<2.2e-16).

Figure 5.1: (a) Cumulative distribution function of in-bubble (echo chambers) and non-bubble (out- and cross-bubble) communication; (b) histogram of distance travelled by messages between sender and receiver in 50-km bins

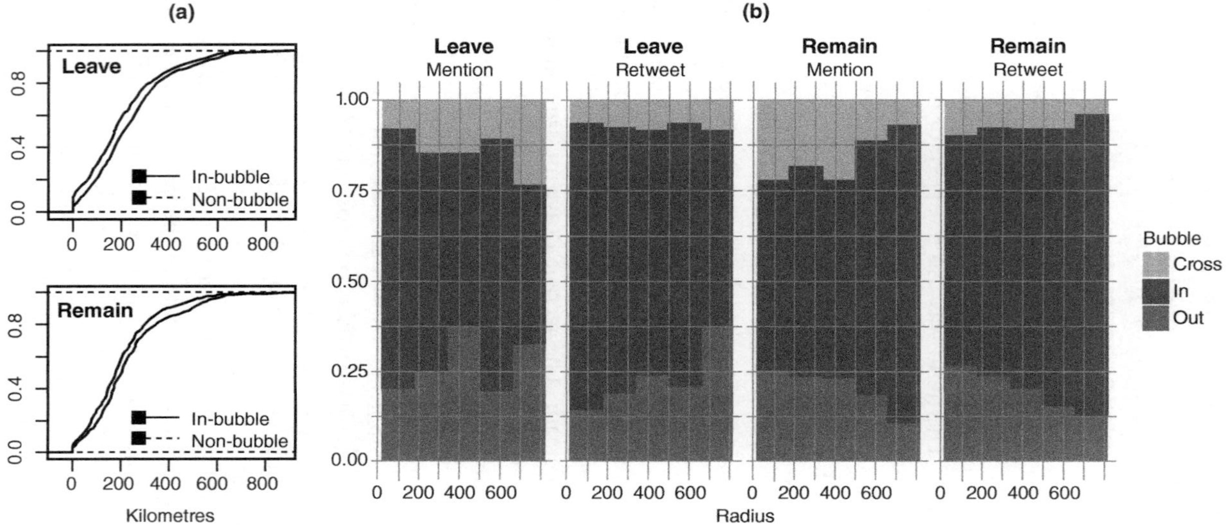

Figure 5.1 unpacks the data and shows the prevailing patterns of echo-chamber communication compared with out-bubble and cross-bubble communication (complementarity), both in the Leave and the Remain campaign, across a range of distance radiuses.

Remain-supporting users were therefore more likely to speak to other Remain supporters outside of their own geographic areas, whereas Leave supporters were largely circumscribed to interaction with users from nearby areas. The differences between echo chambers involving Leave and Remain supporters could be explained by the distinct geographical clustering of their social networks, with communication online echoing the composition of their extant social relations. In summary, while echo chambers in the Leave campaign appear constrained by short geographic distances, Remain echo chambers are likely to span greater geographic distances in contrast to their cross-bubble communication that is physically concentrated around neighbouring communities, an indication that users aligned with the Remain campaign tried to cross the ideological divide in their communities. Figure 5.2 unpacks these geo-spatial social networks.

The robustness checks performed in our study entailed randomly swapping the location of users in each subgraph and recalculating the distance travelled by @-mention and retweet messages. This allowed us to compare the observed distribution of distances against the random distribution of distances travelled by each message. In other words, this approach establishes an association between echo-chamber communication and the geography of message diffusion whenever the observed networks – *ceteris paribus* – differed significantly from the random network. For each iteration of the test, we retain the set of locations in each subgraph, but randomly reorder the locations to test whether geographic dependencies found in echo-chamber communication are replicated in the randomized geographic network. After 100 iterations, we found that the high volume of interactions within

Figure 5.2: (a) Geographic coverage of echo chambers (in-bubble) with number of vertices and edges in each subgraph; (b) central point of diffusion of the Leave campaign in the English Midlands, the North, and the East

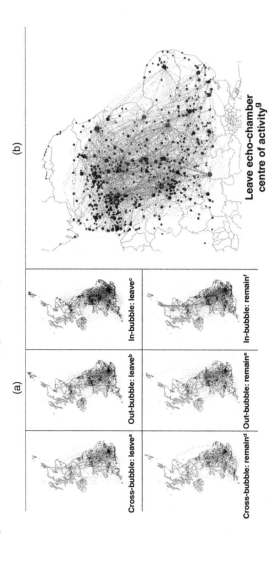

Notes: [a] 1,088 nodes; 1,148 edges; average km covered: 197; [b] 2,853 nodes; 2,772 edges; average km covered: 86; [c] 4,984 nodes; 8,898 edges; average km covered: 22; [d] 972 nodes; 867 edges; average km covered: 243; [e] 1,983 nodes; 1,960 edges; average km covered: 103; [f] 3,184 nodes; 6,015 edges; average km covered: 40; [g] 2,893 nodes; 2,200 edges

geographically proximate echo chambers was a considerable departure from the distribution in the randomized network.

The deviation was particularly prominent in echo-chamber communication; a pattern that disappeared when the location of users was randomly reshuffled. This offered important evidence that the geographic distribution of echo chambers was not determined by chance and allowed us to conclude that the geographic distribution of echo-chamber communication was unlikely, that is, much less likely to happen than in the randomized null model. If anything, the results seemed to suggest that the collapsing of distances brought by internet technologies could foreground the role of geography within one's social network. The unlikely distribution of echo chambers was yet more salient in the subgraph of Leave echo-chamber communication; curiously, it would disappear in non-echo-chamber interactions for the Leave campaign and again in the entire network of Remain interactions. In other words, the association between geographic proximity and echo-chamber communication was particularly salient in the Leave campaign. Figure 5.3 shows the Kolmogorov–Smirnov test for in-bubble, out-bubble, and cross-bubble interaction in the Remain and Leave subgraphs compared with the entire network.

The analysis of echo-chamber communication in the Leave and Remain subgraphs revealed striking interactions between online activity and geography. The results substantiated the existence of geographically bound sociopolitical enclaves materializing in polarized echo-chamber communication online. The results also identified a geographic patterning in online echo chambers, particularly in the Leave campaign, that was exogenous to online interaction. As such, the findings called into question the assumption that echo chambers were a communication effect resulting from online discussion alone. Our expectation was that the unequal geographic distribution of the British population would be observed at each iteration of the tests, with Greater London remaining the central point of information diffusion, and echo chambers reappearing

Figure 5.3: Distances covered by interactions across in-bubble, out-bubble, and cross-bubble for the Leave and Remain campaigns, with Kolmogorov-Smirnov test statistic and p-value

(a) **Network**

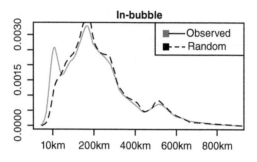

D=0.070274 p<2.20E-16

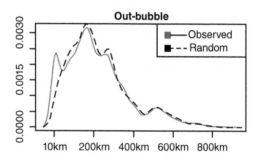

D=0.060904 p<2.20E-16

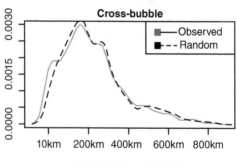

D=0.049132 p<0.01544

Figure 5.3: Distances covered by interactions across in-bubble, out-bubble, and cross-bubble for the Leave and Remain campaigns, with Kolmogorov-Smirnov test statistic and p-value (continued)

(b) **Leave**

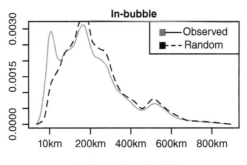

D=0.10665 p<2.20E-16

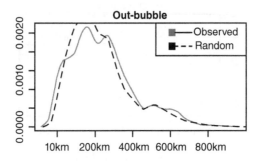

D=0.068182 p=5.07E-06

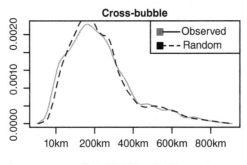

D=0.039199 p=0.341

Figure 5.3: Distances covered by interactions across in-bubble, out-bubble, and cross-bubble for the Leave and Remain campaigns, with Kolmogorov-Smirnov test statistic and p-value (continued)

(c) **Remain**

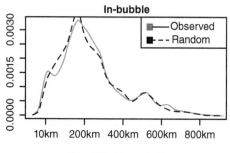

D=0.044223 p=1.56E-05

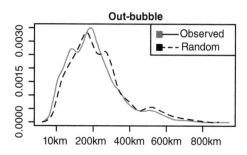

D=0.083673 p=2.20E-06

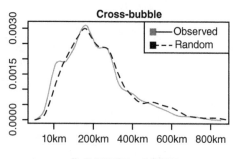

D=0.063437 p=0.06106

Note: The dotted line shows the reference probability distribution used to test the similitude of the two samples with a continuous distribution.

with relatively unchanged geographic coverage. While the network topology remained the same at each iteration of the Kolmogorov–Smirnov test, along with the distribution of users' location, the geographic distribution of echo chambers varied significantly. The bell-shaped, near-normal distribution in the randomized networks is a significant departure from the observed geographic coverage of echo chambers.

The patterns of echo-chamber communication in the Brexit campaign indicate that online echo chambers were likely the result of conversations that spilled over from in-person interactions. As such, they called into question the assumption that echo chambers emerge from social media interaction alone, suggesting instead that users could be bringing their pub conversations to online debate. These differences remained significant even when controlling for outliers such as superusers and abnormalities in the data sliced across weekly series. While the distribution changed over time, the geographic patterning associated with echo chambers in the Leave campaign remained relatively robust, with peak amplitudes that deviate from the rest of the network and from the distribution observed with the random reshuffling of users' locations continuing to appear in the weekly data.

Finally, the differences in echo chambers observed in the Leave and Remain campaigns point to considerable different demographic makeups that underlie their social networks. As such, the significant geographic variation found in the data was likely driven not only by the locations where the two groups were clustered, but also the social positions embedded in the geographical location of Leave and Remain constituencies. Moreover, the different social positions occupied by Leavers and Remainers are consistent with the geographical splintering of the country discussed in the previous chapter. With city-dwellers that voted Remain spending more time shopping or exploring entertainment options outside of their neighbourhoods (Groves, 2006), as well as living and working in hubs of the national and global economy (Storper, 2018), it is unsurprising that Remain groups occupying urban areas would travel more and that their resulting social networks would be more widely connected. The distances covered by their interactions should, therefore, also be lengthier compared with those inhabiting rural or low-density areas of the country, where a higher share of Vote Leave was registered.

SIX

Bots and Trolls

This chapter unpacks the activity patterns of a group of 13,493 bot-like Twitter accounts that tweeted the Brexit referendum and disappeared from the platform shortly after the ballot. The Brexit Botnet comprises 5 per cent of the userbase that tweeted the referendum campaign, a group of users that was removed from the platform by Twitter moderators. The Brexit Botnet sheds considerable light on the weaponization of social media platforms that was central to the referendum campaign, while also exposing the role of algorithms in which bots feature as the most simple, cost-effective, and flexible approach to gaming the social media attention economy. The chapter concludes with a discussion of the policy developments in the aftermath of various congressional and parliamentary inquiries into foreign interference in national elections leveraging bots, trolls, and sockpuppet accounts to weaponize social media.

Bots and superusers

Bots are automatic posting protocols used to relay content in a programmatic fashion. As such, bots are simple algorithms programmed to scrape data from internet sources and post them via social media platforms. Twitter was a relatively bot-friendly platform (Twitter, 2017a) and a number of prominent Twitter accounts were openly bots, including those of established news outlets relaying breaking news, earthquake and tsunami warning systems, or Twitter accounts operated by the Vatican offering regular reflections on Catholic devotion.

In the political sphere, bots can be leveraged to impersonate a third party and are associated with sockpuppets, which are false online identities used to voice opinions and manipulate public opinion while pretending to be another person (Gorwa and Guilbeault, 2018). A large group of bots – a botnet – can be operated as an army of sockpuppet accounts deployed to amplify a defined group of users by retweeting content tweeted by selected users, which may conceivably be bots themselves, a process often referred to as false amplification, chiefly orchestrated by 'fake accounts' (Weedon et al, 2017).

Bots or social bots are, therefore, catch-all terms usually employed to refer to a sockpuppet account. The term sockpuppet draws from the manipulation of hand puppets using a sock and refers to the remote management of online identities to spread misinformation, promote the work of an individual, endorse a given opinion, target individuals, or challenge a community of users (Zheng et al, 2011). Sockpuppet accounts are often automatic posting protocols (bots) operating under a fictitious identity and, as such, they breach the Terms of Service of social networking sites such as Facebook and Twitter. The administration and deployment of bots and sockpuppet accounts are largely centralized and rely on trivial computing routines that allow users and organizations to control substantial subcommunities across social media platforms (Kumar et al, 2017).

Bots are often identified using a combination of metrics, including periods of high-volume posting, followed by a drop in activity levels, sudden deletion at the same time as other suspected bots, activity that does not follow daily human patterns influenced by work and leisure time, high ratio of retweets to tweets, usernames with computer-generated or uncommon words, absence of photo or description in the account profile, high ratio of outward @-mentions to inward @-mentions, high ratio of followers to followees, user account created recently, and low retweet reciprocity (retweeting others, but not being retweeted). The asymmetric data that characterize such accounts stem from their orchestrated behaviour optimized to trigger retweet cascades of selected users in the network with a fraction of the labour required by real users to start cascades of comparable size. Despite the many known characteristics of bots, and the many tools available to inspect specific accounts on demand, it is not easy for real Twitter users to spot bots because the volume of data necessary to recognize their activity patterns often requires systematic data collection.

Although bots rely on trivial computing routines, bot detection is not an exact science and neither human annotators nor machine learning algorithms perform flawlessly (Varol et al, 2017). While human coders are better at generalizing and learning new features from observed data, machine learning algorithms are scalable and regularly outperform human annotators in searching for and detecting complex patterns. Training a machine learning algorithm to identify bots invariably entails a trade-off between recall, the number of correct results divided by the number of possible results, and precision, the ratio of positive and relevant matches, a difficult balance that bot detection algorithms are continuously trying to strike (Rauchfleisch and Kaiser, 2020).

Identifying Twitter bots is therefore a challenging endeavour, as the botmasters can also change tactics to mimic human users and avoid detection. Yet, significant efforts have been made to detect patterns that pertain to bot activity and a growing catalogue of metrics exists for pinpointing political bots (Ratkiewicz et al, 2011b; Abokhodair et al, 2015; Ferrara et al, 2016). While bot detection was originally an enterprise devoted to the identification, demotion, and prevention of spam (Heymann et al, 2007), it has since evolved to mitigate the detrimental impact of malicious activity on electoral politics, policy discussions, and the deliberation of contentious issues. There is compelling evidence that political bots produce systematically more positive content in support of a candidate (Bessi and Ferrara, 2016) and only tweet hashtags used by opponents to disrupt their communication flow (Woolley, 2016). There is also evidence that bots in the 2016 US presidential elections were effective information disseminators (Bessi and Ferrara, 2016). Similarly, during the EU referendum they focused on retweeting content from a selection of users (Howard and Kollanyi, 2016), a marker of their potential to disseminate content.

The literature investigating bot activity is concerned with the imitation of human activity on social media by computer scripts (Bessi and Ferrara, 2016). These algorithms, often referred to as 'social bots', have been shown to approximate (Woolley and Howard, 2016) and upscale human conduct (Bessi and Ferrara, 2016), often influencing communication exchanges on polarizing topics (Howard and Kollanyi, 2016). Social bots can be deployed in a wide variety of contexts and constitute a growing subfield of communication and political science research, which cautions against their harmful impact on electoral politics, policy discussions, and deliberation on contentious issues. Indeed, prominent political events such as the UK EU membership referendum or the 2016 US presidential elections were shown to have been susceptible to such automated interference, especially on Twitter (Bessi and Ferrara, 2016; Howard and Kollanyi, 2016).

Significant efforts have been made to detect patterns of activity that pertain to automation. Evidence to this effect points to the generation and republication of high volumes of partisan content with retweets – the practice of republishing a message already in circulation (Murthy, 2013) – to boost the visibility of said content (Murthy et al, 2016); or, alternatively, to corrupt communication (Woolley, 2016), particularly so as to create 'a false sense of group consensus about a particular idea' (Ratkiewicz et al, 2011b, p 299). Another marker of account automation is the lack of detailed information about the user and the absence of geolocational metadata (Bessi and Ferrara, 2016) that would facilitate detection by users or social platforms (Hwang, Pearce, & Nanis, 2012). Yet, bots can occupy an influential position in communication networks, often appearing at the centre of highly connected network subgraphs in which information diffusion is centralized (Ratkiewicz et al, 2011a).

The investigation by Bessi and Ferrara (2016) into the 2016 US presidential elections found that real users posted more tweets than bots within the period under study. Bots also seemed inept at interacting with humans, replying and @-mentioning primarily other bots.

The study also found that humans tended to respond more frequently to other humans rather than to bots, indicating that humans and bots may operate in largely disconnected subgraphs. These findings differ somewhat from the study of a political botnet by Metaxas and Mustafaraj (2010), which directed replies at recipients purposefully selected for their partisan interest in the Massachusetts elections, a quarter of whom went on to retweet the automated message they received. The various methods of @-mentioning and retweeting suggest that bot operators likely use a range of tactics depending on the political objectives of the botnet.

The bots monitored during the 2016 US presidential elections were nonetheless effective information disseminators. They proved to be effective at spreading information by retweeting content just as frequently as real users. Similarly, during the EU referendum, the most active accounts did not generate new content but rather shared content from other users (Howard and Kollanyi, 2016). While the study acknowledges that humans could achieve comparable levels of activity if they were to focus on retweeting, Bessi and Ferrara (2016) cautioned that bots could hamper human communication precisely because of their relative success at circulating content among real users.

Concerns about the activity of bots and sockpuppets in the context of the Brexit referendum were articulated in the press (Silva, 2016) and academia (Shorey and Howard, 2016), with researchers warning against the automation of political communication and its potential to distort crucial processes at the core of modern liberal democracies, including of course elections (Woolley and Howard, 2016). Howard and Kollanyi (2016) suggested that the EU referendum bots were designed to take a stance on the issue of the UK's membership in the EU. Similarly, Bessi and Ferrara (2016) found that bots produced more positive content in support of a particular candidate, which can skew perceptions of that candidate's support by the public. Similarly, Woolley (2016) argued that accounts exhibiting bot activity play a prominent role in Twitter 'bombs': a barrage of tweets that inundate hashtags used by opponents and are often retweeted by bots to disrupt communication and organization among the opposition.

Howard and Kollanyi (2016) estimated the prevalence of political bots by identifying users that were extremely active in the referendum-related discussion on Twitter. They found that these users were responsible for 32 per cent of all Brexit-related traffic on Twitter. While the study acknowledges that it is difficult to definitively identify which accounts are bots, the authors inferred that the top ten accounts generating the most tweets (over 350 tweets) were likely automated. This is in line with research that has characterized bot activity as 'incessant', which on Twitter translates to 'excessive amounts of tweets' (Bessi and Ferrara, 2016). However, relying solely on user activity to determine the presence of bots is not a reliable method, as individuals can generate numerous tweets by taking turns managing Twitter accounts, a strategy that allows teams to post several hundred tweets per day without any automation (Bastos and Mercea, 2016; Mercea and Bastos, 2016).

The distinction between bots, superusers, and trolls is, however, fraught with challenges. Surpassing bots in complexity and capillarity in the communities they operate, supervised accounts (for example, trolls) were pivotal in the successful disinformation campaign led by the Kremlin-linked and Saint Petersburg-based Internet Research Agency (henceforth, IRA). This campaign relied primarily on supervised accounts operating on Facebook (US District Court, 2018), a sharp contrast to the desolate life of Twitterbots communicating with each other and with modest impact outside their bubbles. The contrast between the two covert strategies raises topical questions about human-driven, curated, and supervised high-volume posting and, conversely, automated, unsupervised, and scripted machine bots. Supervised high-volume posting is a tactic employed by political agents that has received relatively little attention outside of Reddit forums and other online spaces characterized by toxic behaviour (Massanari, 2015).

While bots and trolls continue to be described as political actors undermining democratic legitimacy, there are fundamental differences we first identified in our studies of serial activists, who exhibited extraordinary levels of posting activity combined with a savvy strategy for activating opinion leaders such as journalists, while at the same time assisting activists to coordinate across national boundaries and protest sites (Bastos and Mercea, 2016; Mercea and Bastos, 2016). The observed activity pattern indicated a multi-dimensional engagement strategy that connected online and offline actions across multiple protest sites. This approach was eventually repurposed for sophisticated covert disinformation campaigns, such as those managed by the IRA. These campaigns stimulated partisan communication online and encouraged people to attend rallies in various US states. The IRA also reached out to campaign staff members and appealed to individuals to take their grievances to the streets (Shane and Mazzetti, 2018).

It is against this backdrop that we identified a large network of bots that operated during the Brexit debate (Bastos and Mercea, 2019). We explored the tactics employed by botmasters, deciding which tweets are retweeted and by which subgroup of accounts linked to the botnet. We identified at least two false positives flagged as bots that turned out to be human users, including the very central accounts of @nero, which was operated by the alt-right controversialist and professional troll Milo Yiannopoulos, and @steveemmensUKIP, a Norwich

UKIP and Brexit campaigner. Both accounts were highly connected to the remainder of the botnet and disappeared within the same timeframe, thereby triggering our classifier which relied on thresholding and filtering methods. The number of false negatives in our study most certainly extends beyond these accounts, with similar studies estimating false positives and false negatives in bot detection to hover at around 26 per cent of the data, or 11 and 15 per cent, respectively (Varol et al, 2017). These figures are comparable to the results of our study, as Twitter acknowledges having removed 71 per cent of the accounts identified in our study due to violations of its spam policies (Twitter, 2017b, 2018a).

Bot detection

Bot detection is not an exact science and a range of approaches exist that can be broadly divided between thresholding and filtering methods, with graph-based approaches recently having developed rapidly. The thresholding approach classifies as bots any account that exceeds the detection threshold for individual accounts. This often translates to accounts that tweet above a given number of messages in a defined window of observation. This approach continues to be widely employed in the area (Howard and Kollanyi, 2016) and contrasts to the more selective filtering approach, in which a range of parameters are leveraged to calculate the similarity between Twitter accounts and known bot-like patterns of activity. This is the approach underpinning Botometer (Davis et al, 2016), which quickly became the de facto standard tool for bot detection in internet studies.

The most pressing limitation with Botometer is that it queries Twitter accounts via REST API (Application Programming Interface) and therefore can only inspect active accounts. With Twitter efficiently detecting and removing bots (Twitter, 2017b, 2018), researchers are often unable to use such bot detection algorithms to study account activity retrospectively or to analyse accounts that have been deleted, suspended, or set to private. The solution is to develop a bot detection classifier based on account characteristics known to be associated with automated activity. Such metrics have been detailed by Ferrara et al (2016) and include the number of retweets, tweets, replies, and mentions, but also temporal patterns and username length. Our implementation of a bot detection classifier relied on these inspection steps, in addition to the time series analysis, thus also borrowing from the approach implemented by Abokhodair et al (2015).

As such, we relied on a range of methods reported in the literature of bot detection (Subrahmanian et al, 2016; Varol et al, 2017) to identify a large group of bots whose accounts had been deactivated by the botmaster or blocked/removed by Twitter in the aftermath of the referendum. We relied on the implementation of extended regular expression in R (R Development Core Team, 2014) to identify the campaign associated with tweets and the libcurl implementation (Temple Lang, 2016) to retrieve the webpage title of URLs embedded in tweets (when available).

Previous research has found the frequentist approach to user activity alone to be an unreliable metric to determine the presence of bots, as prolific Twitter posters can tweet abundantly by taking turns and pushing several hundred tweets a day with little to no automation (Bastos and Mercea, 2016; Mercea and Bastos, 2016). Therefore, we analysed multiple measures of user activity along with the posting patterns of potential bots in an effort to distinguish between bots and high-volume posters. This comprehensive analysis enabled us to determine if their activity persisted over time or if there was a significant decline in activity levels that could be considered a 'bot lifecycle' following the EU referendum.

The metrics used to identify bot accounts in our project include detailed profile information, presence or absence of geographical metadata (or propensity to post using web clients), retweet to tweets ratio, @-mention to tweet ratio, activity level, followers to followees ratio, account creation date, and absence of known words in the username. The approach employed to identify automation relied on a combination of filtering and thresholding methods that explored the network structure of retweets and the temporal posting patterns of accounts to classify bots. Positive predictors of bot activity include tweets to user, mean tweet to retweet, common words in the username, use of web interface to relay content, ratio of outbound to inbound @-mentions, ratio of inbound to outbound retweets, account creation date, retweet reciprocity, network transitivity, network modularity, mean and maximum cascade size, share of triggered cascades, and retweet cascade mean time. By modelling the retweet cascade, we managed to identify the retweet networks from bots to users. Figure 6.1 shows the hub-and-spoke formation resulting from the interaction between @nero and bot-like accounts.

Viral content was modelled by rebuilding the retweet cascade. Unfortunately, it is not currently possible to rebuild every step of the retweet cascade, as each retweet includes only a reference to the original message, so that if user C retweets user B who has previously retweeted user A, we can only establish that user A was retweeted by user C, with the

Figure 6.1: Two-tiered botnet, with bots specialized in retweeting active users and bots dedicated to retweeting other bots

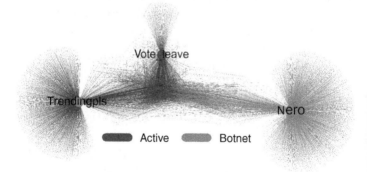

Note: Vertex and edge shade identify source of information.

intermediary steps of the cascade remaining unknown. As such, we could not account for independent entry points that might have influenced the cascade (Cheng et al, 2014). However, given that each retweet includes a unique identifier arranged chronologically from the original tweet to the most recent retweet, we could rebuild cascades from the seed message to the retweets that have cascaded from that original content. Similarly, we relied on the timestamp attached to each tweet to estimate the variable time-to-retweet, calculated as the time elapsed between the original tweet and the ith retweet for cascade of size S.

Brexit Bots

In the third chapter of this book we discussed the political realignment and dealignment that occurred as a result of the EU referendum. This event challenged deeply ingrained ideological beliefs and contributed to a climate of polarization and hyperpartisanship that created favourable conditions for political actors to develop and utilize bots. Campaigners took advantage of these divisions, with the successful Vote Leave campaign coming under criticism for stoking anxiety about immigration and making misleading statements about Turkey's future EU membership. The campaign was also criticized for its disingenuous pledge to boost the failing National Health Service (NHS) by redirecting Great Britain's EU membership contribution into the service (Doherty, 2016; Swinford, 2016). Even though these claims were ultimately shown to be untruthful, it is unclear whether the rectification was of any electoral consequence. Indeed, Vote Leave canvassers continued to effectively rely on data analytics to capitalize on this ostensible tension between the so-called circles of hard-working families and progressive elites (Cummings, 2016), with analyses reporting that social media activity was a positive predictor of the outcome of the vote (Celli et al, 2016).

The sophistication of the operation identified in the Brexit Bots deviates considerably from traditional Twitterbots. Common to most accounts in this botnet was the curated replication of content that was both user-generated and a reproduction of tabloid journalism (Bastos, 2016). Another important marker of this group was the overwhelming prominence of content associated with or authored by user accounts affiliated with the Vote Leave campaign. The overall tone of the messages was much in line with the context of disaffection with immigration and the cultural backlash spearheaded by the older, traditional, and less educated readership of tabloids (Boykoff, 2008). This cultural backlash was strategically leveraged and maximized by populist parties and leaders in order to promote 'traditional cultural values and emphasize nationalistic and xenophobia appeals, rejecting outsiders and upholding old-fashioned gender roles' (Inglehart and Norris, 2016, p 30).

The Brexit Botnet entailed a network of 13,500 Twitter accounts that tweeted extensively about the Brexit referendum, only to disappear shortly after the vote. These Twitterbots posted almost 65,000 messages during a four-week period, with a clear slant towards the Leave campaign. Five accounts in the network alone tweeted 10 per cent of all content posted by bots (@trendingpls, @EuFear, @steveemmensUKIP, @uk5am, and @no_eusssr_thx), with Brexit Bots averaging five tweets a day compared with 1.2 for other users. The study also uncovered two distinct strategies for deploying botnets. A portion of the botnet was dedicated to retweeting other bots while another part only retweeted content from a small set of legitimate user accounts. An example of a particularly active bot was an account named @trendingpls, which tweeted 2,474 messages in the period. Bots were mainly active in the week preceding the vote and on the eve of the referendum, when there was a peak in activity between Twitterbots. Figure 6.2 shows the embeddedness of Brexit Bots in the Twitter debate about the referendum.

Figure 6.2: Activity of Brexit Bots in the weeks leading up to the UK EU membership referendum

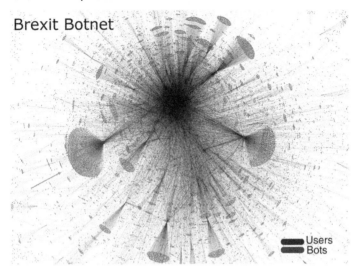

One important contribution of our study was that it sought to measure the impact of bots in the broader Brexit debate. We sought to identify whether bots were used to increase the reach of a given user's message, much like a microphone, or deployed in a concerted effort to amplify a political narrative towards a targeted direction, much like a megaphone. After identifying accounts that both presented bot-like features and that disappeared shortly after the vote, we found that bot-like accounts exhibited clear patterns of specialization that allowed them to trigger small to medium-sized cascades in a fraction of the time required by active users to start cascades of comparable size, but the activity of bots was relatively minor with respect to the larger conversation about the referendum. Our findings indicated that bots can potentially amplify a subset of accounts, but that their influence in the network is limited and falls short of a megaphone.

The content posted by the Brexit Bots was also marked by a markedly short shelf life. This included dubious news stories sourced from a circuit of self-referencing *blews* (Gamon et al, 2008): a combination of far-right weblog such as WorldTribune.com and traditional tabloid media such as express.co.uk. The few webpages still available after the Brexit Bots were removed from Twitter show that the content posted by these users does not conform with the contested and ideologically inflected notion of 'fake news' (Benkler, Faris, Roberts, and Zuckerman, 2017). Instead, the content entailed a form of storytelling that blurs the line between traditional tabloid journalism and user-generated content, which is often anonymous, fact-free, and with a strong emphasis on simplification and spectacularization (Rowe, 2011). User-generated content takes the lion's share of hyperlinks tweeted by these accounts. The content is often presented as a professional-looking newspaper by resorting to content curation services such as paper.li and is likely to include Twitter multimedia (for example, Twitter's native multimedia sharing service twimg.com).

Bots talking to bots

Brexit Bots were effective at triggering retweet cascades and particularly successful at joining ongoing cascades, with a mean cascade participation of $\bar{x} = 18$ (compared with $\bar{x} = 6$ for active users). The botnet was also just as effective at generating large retweet cascades compared with active, regular Twitter users. While regular users started 36 per cent of cascades, the botnet claimed a total of 30 per cent of the retweet cascades. Interestingly, the same long-tailed distribution of hyperactive accounts among regular users is observed in the botnet, where a small share of bots was found to have triggered most retweets, with the remainder of the bots being strategically, although peripherally, positioned to retweet the initial cascade (Figure 6.3). While the group of active users tweeted 97 per cent of the messages and initiated 7.5 per cent of all cascades, the botnet tweeted under 1 per cent of the total messages in the data but accounted for a comparable share of 6 per cent of all cascades.

User activity and cascade time are particularly skewed in the botnet subgraph, with a single user responsible for 4 per cent of all tweets (@trendingpls). But the average cascade time is comparable between the two groups, with botnets starting and completing cascades of size 5, 10, and 20 retweets just one minute faster than active users. The mean cascade time provides an indication that bots mimic the average timespan of retweet cascades, or more likely, that they retweet real users to maximize exposure to the message or to the user posting the original content. Taken together, botnet and the active user groups have not just similar cascade size and time, but also similar averages for cascade mean time at 69 hours for active users and 65 hours for the botnet. For medium-sized cascades ($S=40$, $S=80$, and $S=160$), the botnet completes the cascade 20 minutes faster, but it is with large cascades – $S>320$ and $S<640$ – that larger temporal differences are observed, with the botnet completing such large cascades 1.5–2 hours faster than the active userbase. Figure 6.3 shows this relationship: while cascade time grows linearly along with cascade size for regular users, bots of course face no such constraint.

The Brexit Bots could be neatly divided into two groups or subgraphs. The first group was associated with accounts that leveraged retweet behaviour to amplify the reach of a small set of users and rarely if ever started any cascade themselves. The second group of bots had a narrower scope of operation only retweeting other bots in the botnet, thereby producing many medium-sized cascades that spread significantly faster than the remainder of the cascades. As shown in Figure 6.4, bots that retweeted active users and bots that retweeted other bots exhibit different patterns. As such, the Brexit Bots observed strict and specialized roles in the network, and both groups would eventually feed into the larger pool of real accounts brokering information to @vote_leave, the official Twitter account of the Vote Leave campaign.

Figure 6.4 also shows distinct patterns of human and bot activity, along with interactions that cut across the two groups.

Figure 6.3: (a) Time-to-cascade and (b) mean cascade time for active users and Twitterbots

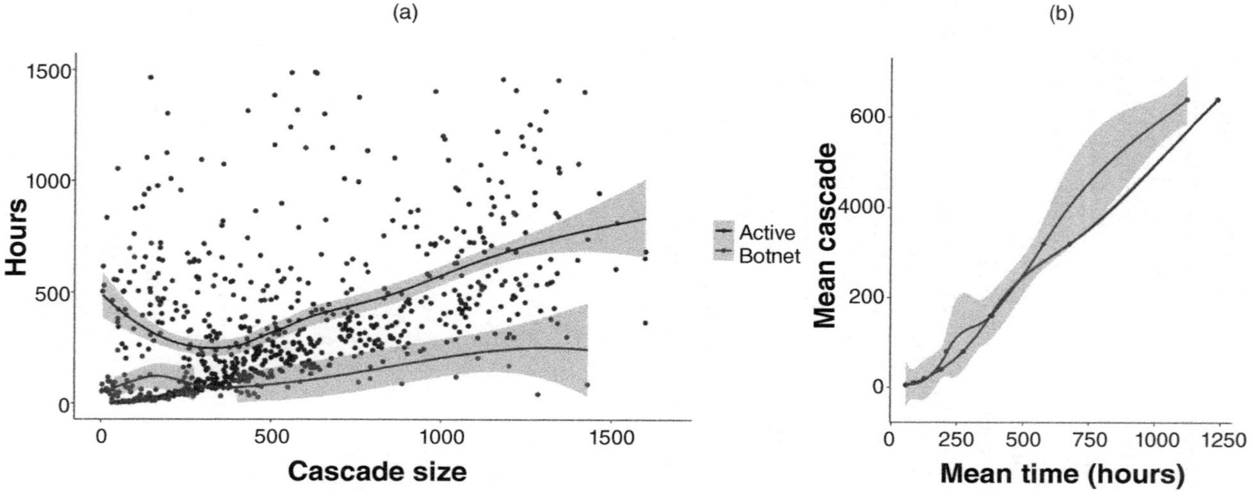

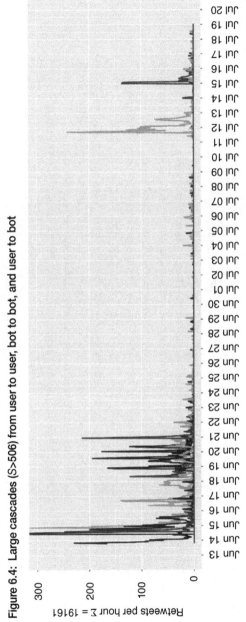

Figure 6.4: Large cascades (S>506) from user to user, bot to bot, and user to bot (continued)

Figure 6.4: Large cascades (S>506) from user to user, bot to bot, and user to bot (continued)

—	USER2USER:	Are you going to #VoteLeave on Thursday? RETWEET thi pledge card to show your support!
—	USER2USER:	Beautiful graphics: Beautiful sentiment #VoteLeave #Brexit
—	BOT2BOT:	Brexit suddenly looking like the smartest thing Britain has done in 50 years
—	USER2USER:	CHECK OUT the video the Eu tried to shut down. This is how the EU spends YOUR money #TakeControl #
—	BOT2BOT:	Haven't said much about #Brexit but I did just leave this on a friend's wall.
—	USER2BOT:	I'm Dutch and I endorse #Brexit because we can be united only in freedom, sovereignty and rule of l
—	USER2BOT:	Londo Muslim Mayor gives #VoteRemain speech. Men at the front, women at the back. Because segrega
—	USER2USER:	RETWEET this pledge card to show your friends that you will #VoteLeave on 23 June! #TakeControl #Pro
—	USER2BOT:	The House of Commons will debate a petition for a second EU referendum on September 5th
—	BOT2BOT:	This is RemainEU Millionaires Jeering at British fishermen #Brexit #VoteLeave
—	BOT2BOT:	Thursday can be 'out country's independence day', Boris Johnson tells #BBCDebate

The figure sheds light on the group of bots that leveraged retweet behaviour to amplify the reach of a small set of users and rarely if ever started any cascade themselves, a strategy that is in sharp contrast to bots whose scope of operation is restricted to retweeting other bots and that would, at times, trigger medium-sized cascades that spread significantly faster. Each of the bot subnetworks played a specialized role and both fed into the larger pool of regular accounts brokering information to the official Twitter account of the Vote Leave campaign (@vote_leave), and arguably the most prominent point of information diffusion associated with the Leave campaign (Figure 6.1).

As for the temporal distribution, their retweet activity was mostly concentrated in the period leading up to the referendum vote (Figure 6.4). Most of it consisted of organic retweets from and to accounts in the active userbase (Figure 6.2). The workflow of bots followed patterns of human use, with each group retweeting active users or other bots, particularly in the week preceding the vote (16–23 June) and on the eve of the referendum (21 June), when we observe a peak in retweet activity between bots. There was a sharp decline in retweet activity after the referendum, primarily among active users who ceased to trigger or join retweet cascades. On the other hand, bots remained operational and activity peaks are observed on 12–15 July: first retweeting active users, then replicating bot content, only to tail off in the following weeks when the botnet was removed from the Twitter platform. In fact, head nodes in the bot network such as @NoThanksEU, @wnwmy, @Foresight1st, @nero, @horrorscreens00, and @Dugher101 disappeared after the end of the referendum (only @NoThanksEU was reactivated in November 2016). This is the critical period (June 2016) when content tweeted by bots and the webpages linked to their tweets disappeared from the internet and from the public Twitter stream altogether.

PART III

Troll Farms Offshore

SEVEN

Information Warfare

This chapter discusses the weaponization of social-media platforms for influence operations centred on the distribution of disinformation and hyperpartisan content, with the Brexit referendum featuring a host of techniques for influence operations that would rapidly spread across local and national elections worldwide. We revisit key events such as the deployment of data-driven microtargeting in political campaigns and the ensuing data lockdown enforced by social media platforms. The chapter covers the literature on disinformation, misinformation, and state propaganda during the Brexit referendum and concludes with a review of the evidence on state and non-state influence operations that sought to strategically diffuse content that heightened partisanship during the Brexit referendum campaign.

Social media propaganda

Propaganda campaigns are often implemented by state actors with the expectation of causing or enhancing information warfare (Linebarger, 1948; Jowett and O'Donnell, 2014). Unlike propaganda targeted at the state's own population, information warfare is waged against foreign states and it is not restricted to periods of armed warfare; instead, these efforts 'commence long before hostilities break out or war is declared ... and continues long after peace treaties have been signed' (Jowett and O'Donnell, 2014, p 212). A key objective of information warfare is to create confusion, disorder, and distrust behind enemy lines (Taylor, 2003; Jowett and O'Donnell, 2014). Through the use of grey or black propaganda, national states in conflict have disseminated rumours and conspiracy theories within enemy territories for morale-sapping, confusion, and disorganizing purposes (Becker, 1949).

Disguised propaganda was a staple of military operations in the 20th century as means of weakening enemy states (Linebarger, 1948), but receded in the aftermath of the Cold War, a period of waning importance for propaganda studies (Briant, 2015). State-sponsored propaganda was largely assumed to rely on mass media such as newspapers, radio, and television (Jowett and O'Donnell, 2014, p 212). During this period, both the definition and forms of propaganda changed dramatically (Welch, 2013), but the centrality of mass media remained a relatively stable component of propaganda diffusion (Cunningham, 2002), a development captured by Ellul (1965) who argued that modern propaganda could not exist without the mass media. Media plurality increased towards the end of the 20th century through the rise of cable TV and the internet, but propaganda operations were seen as a remnant of the past and largely abandoned in scholarly literature (Cunningham, 2002).

The notion that increased media diversity made large-scale propaganda campaigns obsolete continued with the rise of social platforms that enabled citizens and collectives to produce counter-discourses to established norms, practices, and policies. Boler and Nemorin (2013, p 411) reflected this optimism by arguing that 'the proliferating use of social media and communication technologies for purposes of dissent from official government and/or corporate-interest propaganda offers genuine cause for hope'. By the end of the decade, however, this sentiment had changed considerably as the decentralized structure of social media platforms enabled not only public deliberation, but also the dissemination of propaganda. Large-scale actors such as authoritarian states sought to coordinate propaganda campaigns that appeared to derive from within a target population, often unaware of the manipulation (US District Court, 2018). The emergence of social network sites thus challenged the monopoly enjoyed by the mass media (Castells, 2012), but it also offered propagandists a wealth of opportunities to coordinate and organize disinformation campaigns through decentralized and distributed networks (Benkler et al, 2018).

Key events in this change of attitudes towards social platforms include the US presidential election of 2016 and the United Kingdom's referendum on EU membership, where a significant uptake in state-sponsored influence operations was observed. Propaganda studies achieved a new inflection point due to multiple reports of social media platforms being weaponized to spread hyperpartisan content and propaganda (Bessi and Ferrara, 2016; Bastos and Mercea, 2019). These concerns were compounded by the realization that the consolidation and centralization of social platforms allowed state actors to efficiently appropriate social media as channels for propaganda, with authoritarian states seizing the opportunity to enforce mass censorship and surveillance (Youmans and York, 2012; Khamis et al, 2013; King et al, 2017). Technological advances in software development and machine learning enabled automated detection of political dissidents, removal of political criticism, and mass dissemination of government propaganda through social media. These emerging forms of political manipulation also offered extensive anonymity for propagandists (Farkas et al, 2018).

Propaganda models are, however, reminiscent of a media ecosystem dominated by mass media and broadcasting. As such, the classic propaganda models probe into processes of framing, priming, and schemata, along with a range of media effects underpinning information diffusion in the post-war period leading up to the Cold War (Hollander, 1972), but invariable predating the internet (Hermans et al, 2015). Propaganda classes are divided into white, black, and grey. White propaganda refers to unambiguous, openly identifiable sources, in sharp contrast to black propaganda in which the source is disguised, with grey propaganda sitting somewhere in between these two (Becker, 1949; Doherty, 1994; McAndrew, 2017). The influence operations identified in the Brexit debate typically entail a case of black propaganda, as the source was a state actor disguising the operations as domestic. There are instances, however, of propaganda that appeared to have been sourced locally in Britain, in which case the Brexit influence operation is classified as white propaganda given the reliance on openly identifiable domestic sources.

In the following sections we unpack the black propaganda campaign during the Brexit referendum led by the Internet Research Agency (IRA), a so-called 'troll factory' that was also known as the 'trolls from Olgino' or 'Kremlin bots', and whose operations were reportedly linked to the Russian government (Bertrand, 2017). The IRA was dedicated to disseminating propaganda and influence operations on behalf of Russian business and political interests. The organization managing the 'troll factory' and the Kremlin bots was reportedly created by Yevgeny Prigozhin, a former Russian oligarch who also emerged as the leader of the mercenary company Wagner Group, whose military operations were similarly international, having operated extensively in Syria, Libya, Africa, and most notably in Ukraine during the Russo-Ukrainian war.

We identify such 'Kremlin bots' by relying on a list of deleted Twitter accounts that was handed over to the US Congress by Twitter on 31 October 2017 as part of their investigation into Russia's meddling in the 2016 US elections (Fiegerman and Byers, 2017). According to Twitter, a total of 36,746 Russian accounts produced approximately 1.4 million tweets in connection to the US elections (Bertrand, 2017). Out of these accounts, Twitter established that 2,752 were operated by the IRA (United States Senate Committee, 2017). In January 2018, this list was expanded to include 3,814 IRA-linked accounts (Twitter, 2018d). We conclude this chapter with a review of the content posted by the Brexit Bots, which unlike the IRA trolls were reportedly based in Bristol and would therefore constitute white propaganda.

The IRA troll farm

The IRA is a secretive private company based in St Petersburg, Russia, reportedly orchestrating subversive political social media activities in multiple European countries and the US, including the 2016 US elections (Bugorkova, 2015; *The Economist*, 2018). The US District Court (2018) concluded that the company engaged in information warfare based on 'fictitious US personas on social media platforms and other Internet-based media'. The court also linked the IRA to the Kremlin through its parent company, which holds various contracts with the Russian government. There is also evidence linking the founder of the IRA, Yevgeniy Prigozhin, to the Russian political elite, and of course to the Wagner Group that would ultimately seal the fate of the oligarch. The Russian government has nonetheless rejected accusations of involvement in subversive social media activities and downplayed the US indictment of Russian individuals (MacFarquhar, 2018).

The IRA has been dubbed a 'troll factory' due to its engagement in social media trolling and the incitement of political discord using fake identities (Bennett and Livingston, 2018). This term has clear shortcomings, as the agency's work extends beyond trolling and includes large-scale subversive operations. According to internal documents leaked in the aftermath of the US election, the workload of IRA employees was rigorous and demanding. Employees worked 12-hour shifts and were expected to manage at least six Facebook fake profiles and ten Twitter fake accounts. These accounts produced a minimum of three Facebook posts and 50 tweets a day (Seddon, 2014). Additional reports on the subversive operations of the IRA described employees writing hundreds of Facebook comments a day and maintaining several blogs

(Bugorkova, 2015), largely aimed at sowing discord among the public.

The Twitter profiles operated by the IRA combined human-operated and automated accounts to perform as de facto crowd-sourced elites that occupied central positions in retweet networks throughout 2016 (Badawy et al, 2019). IRA operatives engaged in trolling by intervening in an online conversation to spark a reaction among readers at the behest and for the benefit of their patron. Leveraging black, white, and grey propaganda, IRA activity likewise testified to the aim to amplify a resonant discourse among targeted subcultures through the impersonation of its membership (Stewart et al, 2017; Bastos and Farkas, 2019; Freelon et al, 2020). In contrast to traditional public influence propaganda, marked by mass produced, mass distributed campaigns, the IRA activity was remarkable in its ability to identify and appropriate the subculture or social identity of relatively narrowly defined targets (Jensen, 2018; Kim et al, 2018; Linvill and Warren, 2020).

Previous research focused on influence operations conducted by the IRA found that the agency sought to spread content on both sides of the partisan divide. The researchers analysed IRA profiles and found that they were constructed as impersonations of imagined audiences (Arif et al, 2018), which on Twitter entail both direct connections with followers and indirect links with the larger Twitter social graph through @-mentions or retweets (Marwick and Boyd, 2011). These profiles were tailored to specific partisan groups, a sophisticated operation that allowed the agency to present their operatives as authentic and interesting users (Arif et al, 2018). The IRA curated a set of accounts whose profile information and pictures self-describe themselves as British, African American, Conservative, or simply as young women, demonstrating cultural acuity and sensitivity to the social identity of supporters and detractors of the targeted movement on Twitter (Bastos et al, 2021a). By doing so, the IRA was able to show an understanding of the values, symbols, and views associated with different group memberships (Tajfel, 1978; McGarty et al, 2014).

The IRA Brexit campaign

The Twitter Moderation Research Consortium (formerly Elections Integrity, 2018) identified a group of accounts operated by the IRA and subsequently banned from the platform. The first release of the data included 2,752 accounts the company recognized were operated by the IRA (United States Senate Committee, 2017). This list was expanded in early 2018 to include 3,814 IRA-linked accounts (Twitter, 2018d). The data set was updated again to include 3,836 Twitter accounts, along with the totality of images, videos, and profile images used or created by these accounts. Although some accounts were created as far back as 2009, most were created in 2013 (29 per cent) and 2014 (40 per cent), with a smaller share of accounts created in 2015 (12 per cent) and 2017 (12 per cent). These accounts amassed a combined following of over 6 million users ($N=6,386,749$).

A subset of the tweets posted by these accounts was manually and systematically annotated by an expert coder to identify the most prominent issues mentioned by the account (Farkas and Bastos, 2018; Bastos and Farkas, 2019). Each IRA account was coded based on three variables: user type, national identity, and campaign target. Campaign target was established by training a set of 250 accounts to render a typology of campaign targets of IRA-operated accounts in our database. The typology was created based on recurrent identifiers in account descriptions, language, time zone, nationality, and tweeted content. We ultimately identified several distinct target groups for the influence operation, one of them being Brexit. These groups are: *Russian citizens* (including Russian politics, Russian news, and self-declared Russian propagandists); *Conservative patriots* (including Republican content); *Protest activism* (including Black Lives Matter, Anti-Trump, and Anti-Hillary communication); *Local news* (whose accounts mostly post and retweet a curated set of mainstream media sources); and finally *Brexit* (including mainstream media coverage and support to the Brexit campaign).

The Brexit content posted by the IRA trolls comprise typical black propaganda, as these accounts presented themselves as British, European, or American when they were, in fact, operated by Russian operatives. The content they posted was often emotionally charged and trafficked on fearmongering, populist sentiment, polarization, hostility, and conspiracy-theorizing. It often endorsed the Brexit campaign or the supporting organizations (for example, Leave.EU and Vote Leave), or expressed disapproval of adversarial organizations. It also shared rumour and conspiracy theories, along with the occasional encouragement of action (for example, 'vote X' or 'share this!'). Populist rhetoric was rampant, with several references to 'the people', 'anti-establishment', and 'anti-mainstream media'. It also featured scapegoating, ethno-cultural antagonism, and narratives that foregrounded perceived threats to society and the need for a strong leader.

By applying a coding typology to the data based on target campaigns, we also found that grey propaganda was significantly predictive of content targeting local news and the Brexit campaign, whereas white propaganda was unsurprisingly associated with Russian and Ukrainian campaigns. However, 75 per cent of the content targeting the Brexit campaign was classified as black propaganda, as the accounts that posted the content disguised their location and real affiliation. Only 5 per cent of Brexit-related content crafted by the IRA was posted by accounts publicly identified as being Russian or affiliated with Russian organizations.

The year 2013 marks the inception of the black propaganda operations that were pivotal to the IRA involvement with the Brexit campaign, with over one-quarter of such accounts created in this period. The temporal patterns identified in these operations are consistent with the strategic objectives of the campaign. One-quarter of the Twitter accounts created by the IRA and employed in the Brexit campaign were registered between 2013 and 2015, whereas half of the entire universe of IRA Brexit accounts were registered only in the run-up to the 2016 Brexit campaign. Indeed, 2016 alone accounts for 84 per cent of the activity tweeted by these accounts. In other words, the IRA Twitter accounts dedicated to Brexit remained largely dormant until 2015 and early 2016, the period leading up to the referendum when 80 per cent of their tweets were posted.

A significant uptake in the creation of black propaganda was also observed in the following year (2017), but their activity decreased, likely due to Twitter terminating this network of sockpuppet accounts seeding propaganda. Grey propaganda accounts, which also posted Brexit content, appear to be the most complex operation carried out by the IRA. One-third of these accounts were created in 2013 and a further 42 per cent in 2014. While 83 per cent of grey accounts were created before 2014, they remained largely dormant until 2016, when half of the messages tweeted by these accounts were posted. Indeed, the median activity of grey accounts fell on 29 June 2016, which was just one week after the UK EU membership referendum. As such, much of Brexit influence operation manned by the IRA campaign followed a short-term organizational pattern similar to white propaganda, with a considerable portion of the accounts being registered only a few months before the referendum vote.

Troll farms in Bristol

The Brexit campaign was also marked by domestic influence operations that could be classified as white propaganda. This is consistent with the Leave.EU response to the reportage of James Ball for Buzzfeed on our research uncovering the Brexit Botnet (Bastos and Mercea, 2019). On that occasion, Andy Wigmore, the head of communications for Leave.EU during the referendum campaign, revealed they used a 'creepy'

Facebook profiling technology to target voters with anti-EU messaging. The Leave.EU campaigner also tweeted that 'we had our own bots in Bristol and we used Artificial Intelligence to target specific groups—it worked because we knew whom to hit' (Miller, 2017).

The admission of the Leave.EU campaigner that their bots were domestic, after all they were based in Bristol, was a response to pressure to clarify whether the Brexit Bots were, in fact, affiliated with the Russian outfit known as the IRA. A closer inspection of this pool of accounts indicates that slogans associated with the Vote Leave campaign were used in a ratio of 8:1 compared with Remain slogans. The Brexit Bots were also significantly more active in the period leading up to the referendum, with an average of 4.4 messages compared to 3.9 for the rest of the population ($\bar{x} = 4.44\ \sigma = 33.3$, and $\bar{x} = 3.99\ \sigma = 74.2$, respectively), and significantly less so in the wake of the vote, with an average of 2.4 tweets compared to 2.6 for the global population ($\bar{x} = 2.42\ \sigma = 9.0$, and $\bar{x} = 2.61\ \sigma = 63.2$, respectively).

A closer inspection of the content posted by the Brexit Bots also allows for examining the intersections between Twitter activity and the broader informational environment around the platform. This was achieved by assessing the circulation and amplification of media objects from outside sources (videos, images, weblinks, and so on) in the tweets. Successful influence operations operate across platforms to maximize the attention mechanism built into our high-choice media ecosystem, and the cross-posting, cross-linking performed by this cohort of accounts reveals the role of Twitter in disseminating partisan views during the Brexit debate. As such, we examined the dissemination of external media objects (YouTube videos, image memes, news URLs, and so on) to position Twitter in relation to the broader media ecology in which the platform is situated.

Much of the content posted by the Brexit Bots has unfortunately vanished from the internet. Upon attempting to resolve the links tweeted by this cohort of accounts, we found that the majority of tweeted URLs (55 per cent) no longer exist, cannot be resolved, or link to either a Twitter account or a webpage that no longer exists. Nearly one-third (29 per cent) of the URLs link to Twitter statuses, pictures, or other multimedia content that is no longer available and whose original posting account has also been deleted or blocked, a marker of the perishable nature of digital content at the centre of political issues (Walker, 2015). From this total, a full 1 per cent of the links are directed to user @brndstr, one of the few accounts appearing in the communication network of the Brexit Bots that remained active long after the referendum. This account is managed by the Dubai-based 'Bot Studio for Brands', a company specialized in providing Twitter bots for social media campaigns, whose Twitter account was later suspended by Twitter and then recreated in 2020.

The few links that remained accessible eight months after the referendum can hardly be described as fake news, at least as long as fake news is defined as intentional, misleading half-truths and/or outright lies (Farkas and Schou, 2019). On the other hand, the material is rich in rumours, unconfirmed events, and human-interest stories with an emotional and populist appeal that is typical of tabloid journalism (Bastos, 2019). As discussed in the previous chapter, the sources we managed to inspect, though not representative of the much larger universe of content tweeted by this population of users that vanished from the Twittersphere, is much akin to hyperpartisan tabloid journalism, with a topical emphasis on highly clickable, shareable, and human-interest driven stories rarely covering any objective news event. Table 7.1 unpacks and summarizes the URLs tweeted by the Brexit Bots.

Although 17 per cent of weblinks direct to Twitter accounts that are still active and reachable, a random sample shows that the original status is often no longer available, thus preventing any determination of the content originally tweeted. A good

Table 7.1: Weblinks tweeted by the Brexit Bots

1.	*Dead link (external sources)*	54.30%	11.	facebook.com	0.50%
2.	*Valid link (Twitter)*	17.20%	12.	dailymail.co.uk	0.50%
3.	*Dead link (Twitter)*	8.60%	13.	amp.twimg.com	0.50%
4.	express.co.uk	1.70%	14.	*Suspended account*	0.30%
5.	theguardian.com	1.60%	15.	edition.cnn.com	0.20%
6.	youtube.com	1.40%	16.	petition.parliament.uk	0.20%
7.	bbc.co.uk	1.10%	17.	virgin.com	0.20%
8.	@brndstr account	1.00%	18.	paper.li	0.20%
9.	telegraph.co.uk	0.60%	19.	thesun.co.uk	0.20%
10.	bloomberg.com	0.50%	20.	reuters.com	0.20%

example is the tweet ID 740138870092750848 posted by user @evangeliney0ung with several hundred retweets. Although the user remains active, the original retweet has been removed (together with the relevant retweet cascade), and with Internet Archive having no record for this specific tweet, it is no longer possible to know what the original photo conveyed. The high deletion rate, not only of accounts but also of weblinks tweeted by this population, suggests that the Brexit Bots were devised for a short-term campaign to which the permanence of the content over time was irrelevant.

The Brexit Bots also linked to media sources both mainstream and user generated. In addition to the many dubious news stories sourced from a range of self-referencing *blews* (Gamon et al, 2008), which include blogs and tabloid media such as express.co.uk, they also linked to mainstream media that was either quoted out of context, framed to emphasize a partisan aspect of the reportage, or edited with incomplete facts so that it suppressed any opposition-idea, including reporting by Sky News, an authoritative source of journalism even if it is associated with bringing US 'helicopter journalism' to the UK (Lewis et al, 2005). The original reportage aired by Sky News features an interview where the reporter asks 'the woman on the street' if she had 'tried to get a job and found that it has gone to somebody else?' This snapshot of an otherwise legitimate source of journalism featured prominently in the Brexit Botnet, and it rarely featured the response of the interviewee clarifying that, in fact, she had not.

EIGHT

Social Media Manipulation

This chapter discusses the emergence of the disinformation landscape that supported influence operations deployed in the Brexit referendum, a watershed development resulting from the shift in the governance of social technologies from communities of users to social media algorithms. We revisit key events such as the deployment of data-driven microtargeting in political campaigns epitomized by the Cambridge Analytica data scandal and the ensuing data lockdown enforced by social media platforms. These developments severely restricted independent research on mis/disinformation campaigns by preventing data access to social media data and rendering problematic content, including mis- and disinformation, increasingly more inscrutable and unobservable. The chapter concludes with a set of lessons learned from studying disinformation during the Brexit campaign.

Networked publics to social platforms

Much of the policy intervention targeting social media that followed the Brexit debate stems from the conclusion that misinformation and disinformation pose a serious threat to objective decision-making by the voting public (Lewandowsky et al, 2017). The success of mis- and disinformation campaigns has been partially attributed to strategies that maximize the biases intrinsic to social media platforms (Comor, 2001; Innis, 2008; Benkler et al, 2018), particularly the attention economy and the social media supply chain that relies on viral content

(Jenkins et al, 2012) for revenue generation. This backdrop of influence operations and information warfare presents a considerable departure from years of euphoric rhetoric praising the democratization of public discourse brought about by networking technology and social media platforms (Howard and Hussain, 2013).

Early scholarship that heralded the potential of social media for democratization and deliberation inadvertently reinforced a narrative championing communication and collaboration as expected affordances of social platforms (Shirky, 2008). By the end of the decade, however, the narrative surrounding social media platforms increasingly turned to metaphors foregrounding polarization and division in a landscape marked by tribalism and information warfare (Bastos, 2021a, 2022b), enabled by a business model driven above all by the commodification of user trace data and the integrated dependencies connecting big tech to financial markets (Langley and Leyshon, 2017).

Scholarship on this hybrid media ecosystem (Chadwick, 2013) explored the technological affordances and ideological leanings that shape social media interaction, with a topical interest in the potential for civic engagement and democratic revitalization (Zuckerman, 2014). Bennett and Segerberg (2013) expanded on Olson's seminal work on the logic of collective action to explain the rise of digital networked politics where individuals would come together to address common problems. Similarly, Castells (2009, 2012) described a global media ecology of self-publication and scalable mobilization that advanced internet use and political participation.

Open platforms and unrestricted access offered the cornerstone of networked publics that reconfigured sociality and public life (Boyd, 2008). The relatively open infrastructure of networked publics was also explored in scholarship detailing how online social networks support gatewatching (Bruns, 2005) and practices in citizen journalism that are central to a diverse media ecosystem (Hermida, 2010), with citizens auditing the gatekeeping power of mainstream media and holding elite interests to account (Tufekci and Wilson, 2012). By most assessments, social network sites were welcoming challengers to the monopoly enjoyed by the mass media (Castells, 2012), with only limited attention devoted to the opportunities offered to propagandists that could similarly coordinate and organize disinformation campaigns through decentralized and distributed networks (Farkas and Schou, 2019).

Colonized by algorithms

These developments challenged the very idea of networked publics and Castells' (2012) account of the internet as universal commons. The transition from narratives emphasizing open communication to concerns about social media manipulation warfare was accompanied by the steady transition from networked publics, centred on a user-centric and decentralized governance framework, to algorithmic-driven commercial platforms. These services built much of their social infrastructure on the back of networked publics and the community organization that shaped internet services in the early 1990s. In the early 2010s, social platforms consolidated their grip on the social infrastructure by replacing desktop-based applications with mobile platforms, a transition that substituted open standards with cloud-based, centralized application interfaces controlled by social media platforms (Walker, Mercea, and Bastos, 2019). By the early 2010s, the colonization of user communities was completed through the large-scale implementation of AI systems that managed interactions and attention, including Facebook's News Feed and Twitter's Trending Topics. These changes mirrored a corporate transition of social platforms as business centred on services to the leasing and trading of user data.

Also noticeable in the transition from networked publics to social platforms was the increased commercialization of previously public, open, and often collaborative spaces, promptly transformed into corporate owned, walled-off platforms. Social media platforms became centralized gatekeepers to critical infrastructure supporting economic, democratic, and social participation. Operating in an open market with limited regulation or external oversight, social platforms flourished in an environment that supported the continuous upscaling of social infrastructure with no public-facing system of governance. This change in the social infrastructure came to public attention in the aftermath of the Cambridge Analytica data scandal, when the personal information of millions of Facebook users was collected for political advertising without their consent. The profitable collection of user data at scale and speed triggered a public outcry on digital privacy, data access, surveillance, microtargeting, and the growing influence of algorithms in society. It also revived efforts to implement distributed networks services such as Mastodon or Pleroma.

As such, social platforms built much of their social infrastructure on the back of networked publics and the community organization that shaped internet services in the early 1990s. Indeed, the drive towards community formation remains a key component of social media platforms, notwithstanding the growing trend towards data access restrictions, end-to-end encryption of messaging apps, and virtual reality environments such as the Facebook Metaverse. In February 2017, at a time when investigations into disruptive communication in the previous year's US elections were still evolving, the CEO of Meta (then Facebook) Mark Zuckerberg doubled-down on the global community mission of the company by emphasizing that their mission was 'to build social infrastructure like communities. ... Progress now requires humanity coming together not just as cities or nations, but also as a global community' (Zuckerberg, 2017).

The algorithmization of communities championed by social platforms was a milestone that instantaneously rendered networked publics into a profitable source of users' interactions. Transferring community governance from users to algorithms removed a key basis for mutual trust, opening the way for large-scale disinformation campaigns that conspicuously plagued election cycles, ethnic relations, and civic mobilization from 2016 onwards (Apuzzo and Santariano, 2019). By Facebook's own account (Weedon et al, 2017), its advertising algorithms were harnessed to segment users into belief communities that could be micro-targeted with materials that amplified their intimate political preferences. This repurposing of intimate knowledge and networked interaction for revenue-making remained the corollary of commercial social media enterprises, including the individuals and academics involved in the notorious and now defunct political consultancy firm Cambridge Analytica (Rosenberg, 2018).

Gaslighted by social platforms

The lockdown of social platforms' Application Programming Interfaces (APIs), especially that of Facebook and Instagram in the wake of the Cambridge Analytica data scandal and the congressional hearings post-2016, has hindered research on influence operations and disinformation in meaningful ways. Mitigation strategies available to the public and the academic community are inadequate because independent source attribution is near impossible in the absence of digital forensics (Bastos, 2022a). The monitoring tools made available after the lockdown of APIs, including data access facilitated by Social Science One and CrowdTangle, which is owned by Facebook, are fundamentally imperfect because no direct access to data is possible. Similarly, while Twitter has offered archives of disinformation campaigns that the company identified and removed (Elections Integrity, 2018), such sanctioned archives offered only a partial glimpse into the extent of influence operations and may prevent researchers from examining organic contexts of manipulation (Acker and Donovan, 2019).

With no access to Facebook and Instagram data, arguably the most important platforms for propaganda and influence operations, independent source attribution and the monitoring of disinformation is near impossible. As such, our understanding of what constitutes mis- and disinformation and how widespread the problem is on social platforms is tied to, and depends on, the fragmented data that platforms such as Twitter and Facebook release to limited segments of the academic community, usually research institutions that have struck an agreement with the companies. This rather limited sample of disinformation campaigns effectively shapes our understanding of what strategies are in place, how large these networks of disinformation are, and what strategies of mitigation can be employed to control the spread of problematic content (Bastos, 2022a).

The available data are in no way representative of various disinformation campaigns occurring on the platforms. It is often the case that influence operations are identified by researchers and journalists unaffiliated with the social platforms, so no one has a complete picture of the strategies taking place online at any given time. Even disinformation circulating on public platforms such as Twitter can only be detected to a limited extent. This is because Twitter's Terms of Service (ToS) state that content deleted by a user or blocked by the platform due to infringements on the ToS ought to disappear from the platform altogether (Twitter, 2019a). Similarly, deleting a tweet automatically triggers a cascade of deletions for all retweets of that tweet (Twitter, 2018b). This specific affordance of social platforms facilitates the disappearance of posts, images, and weblinks from the public view, with lasting effects on research on influence operations.

The solution to this problem that sits at the centre of our disinformation landscape is to provide data access and oversight of the interactions that take place in social platforms. Public and open APIs allow researchers to retrieve large-scale data and curate databases associated with contentious and sociologically meaningful events. Without them, web interfaces have to be scraped to access the data (Freelon, 2018), which is labour intensive and drastically limits the amount of information that can be collected and processed. Locking researchers out of the APIs constrains them to human-intensive means of data collection that cannot produce large or representative samples of real-world events, such as social movements, elections, and of course state and non-state sponsored disinformation campaigns.

Illustratively, Twitter used to operate three well-documented public APIs in addition to its premium and enterprise offerings (Twitter, 2019a). Twitter's relative accessibility, now in peril following the acquisition of the company by Elon Musk, leads it to be vastly over-represented in social media research (Blank, 2017). Public and open APIs such as that of Twitter are an exception in the social media ecosystem. By contrast, Facebook's Public Feed API (Facebook, 2018b) is restricted to a limited set of media publishers. Alternative methods to data access cannot match the extensive and scalable data access provided by APIs, as these remain integral to mobile and cloud-based technologies that are central to the infrastructure of social media platforms. With the ToS of social platforms often preventing researchers from sharing data, including Twitter, APIs allow for the programmatic retrieval of each post's unique identification number, which researchers can then share to ensure reproducible research.

The disinformation landscape

The current disinformation landscape was marked by landmark influence operations in elections worldwide, with prominent examples including the 2016 US elections and the 2017 general elections in France (Bessi and Ferrara, 2016; Ferrara, 2017; Weedon, et al, 2017). These developments required the adoption of specialized vocabulary associated with disruptive operations to describe a set of media practices designed to exploit deep-seated tensions in liberal democracies (Bennett and Livingston, 2018). These tactics are optimized to polarize voters and alienate them from the electoral process (Benkler et al, 2018), a playbook infamously associated with the Russian trolls of the Internet Research Agency (IRA) discussed in the previous chapter. They maximize the strategic and political use of inaccurate information, that is, misinformation (Karlova and Fisher, 2013), and disinformation, or the intentional distribution of fabricated stories to advance political goals (Bennett and Livingston, 2018).

This new vocabulary projected the growing realization that the public is largely unaware of the multiple influence operations breeding at the intersection of social platforms and partisan politics. With social platforms increasingly organizing, filtering, and displaying information based on algorithms that are largely inscrutable and unobservable even for those working in this corporate environment, the experience of using social media is marked by an intangible sense of alienation and manipulation. This sense of broad and unqualified manipulation that is difficult to pinpoint has effectively fed much of the rampant conspiracy theorizing that emerged in the 2010s and blossomed in the following decade, including QAnon to anti-vax narratives. Disinformation campaigns remain similarly elusive, with social platforms playing a cat-and-mouse game with 'bad actors' who, for the most part, can effortlessly evade prosecution.

There are also technical constraints that curb mitigation strategies to disinformation campaigns. Resource allocation for the detection and removal of influence operations is notoriously time consuming, whereas the time involved in planning and execution of disinformation campaigns is typically short. Influence operations can thus leverage the 'firehose of falsehood' model (Paul and Matthews, 2016) whereby a large number of messages are broadcast rapidly, repetitively, and continuously over multiple channels without commitment to consistency or accuracy (Bertolin, 2015). Content flagged by a social platform's algorithms and partnering fact-checking agencies is quickly removed, so that disinformation is phased out and disappears as soon as rectifying information emerges.

But disinformation campaigns can quickly republish the same content, whereas the mitigation strategies employed by social platforms require extensive tracking of information in real time for limited results.

This situation has forced researchers and journalists monitoring disinformation campaigns to work with fragmentary evidence and second-guess the algorithmic decisions that resulted in the purging or downranking of content. The reverse engineering of social platforms, commonly referred to as 'algorithmic auditing', requires extensive digging into disinformation as it happens in real time and with limited support from social platforms. Even when individual users and journalists report potential disinformation campaigns, social platforms rarely disclose content that was flagged for removal, and therefore studying influence operations on social media becomes an exercise in reverse engineering at multiple levels, with the most prominent being the interplay between the strategies and intentions of malicious state and non-state actors and the limited amount of evidence (data) made available by social platforms. This has severely hindered the identification of influence operations in real time, which is currently carried out only retrospectively, after large influence operations have already inflicted damage.

Current systems that can prevent harm on social media are, therefore, severely inadequate, as influence operations can routinely daisy chain multiple harassment and disinformation campaigns that are phased out and disappear as soon as rectifying information or alternative stories start to emerge. The low persistence and high ephemerality of social media posts are leveraged to transition from one contentious and unverified frame to the next before mechanisms for checking and correcting false information are in place. As such, influence operations can easily exploit the opaqueness and inscrutability of social platforms by offloading problematic content that is removed from platforms before the relentless – but ultimately time consuming – news cycle has successfully corrected the narratives championed by highly volatile social media content.

The absence of accountability and oversight mechanisms for social platforms, and a context where influence operations can easily leverage the firehose of falsehood, maximize the vulnerability of those targeted by disinformation. Individuals find themselves unable to tell whether mass harassment and brigading are coordinated or not, and the decision-making process regarding content that has been reported or flagged for removal is restricted to social platforms' content policy team, who decides on individual cases with little to no external input. The opaqueness and the politics of deletion implemented by social platforms is beneficial to influence operations because disinformation performs well in short timeframes. Even when content is routinely removed, the high-volume posting is effective because users continue to be exposed to content removed from social media through reposting and resharing from multiple sources.

PART IV

Politics Erased

NINE

The Politics of Deletion

This chapter takes stock of the remarkable and disturbing rate of deletion of social media content by considering Brexit data that are now unavailable. We unpack the politics of deletion by inspecting the ratio of deleted tweets and accounts that participated in the Brexit debate. We show that 33 per cent of the tweets leading up to the referendum vote have been removed, that only about half of the most active users who tweeted the referendum continue to operate publicly, that 20 per cent of the accounts that tweeted the referendum are no longer active, and that more messages from the Leave campaign disappeared than made up the entire universe of tweets affiliated with the Remain campaign. We argue that while the ephemerality of social media posts may be a reasonable expectation, the magnitude of content deletion poses considerable challenges for informed public deliberation around matters where the issue being deliberated on is constantly disappearing from public scrutiny.

Politics erased

Influence operations and propaganda on social media emerged in the run-up to electoral events in 2016 and continue to challenge policy makers and researchers. These operations rely on coordinated and targeted attacks where the accounts and profiles sourcing the content disappear in the months following the campaign. User accounts may be suspended from social platforms for violating standards and Terms of Service (ToS), such as posting inappropriate content or displaying bot-like activity patterns; others are deleted by the malicious account holders to cover their tracks (Owen, 2019). The modus operandi of influence operations often consists of amplifying original hyperpartisan content by large botnets, as we detailed in Chapter Six, that disappear after the campaign. The emerging thread is then picked up by high-profile partisan accounts that seed divisive rhetoric to larger networks of partisan users and automated accounts (Bastos and Mercea, 2019).

It is against this landscape of information warfare that political campaigns seek to influence public opinion. Social media platforms ramped up efforts to flag false amplification (Weedon et al, 2017), remove 'fake accounts' (Twitter, 2018b), and prevent the use of highly optimized and targeted political messages on users (Dorsey, 2019). These efforts sought to clear social platforms of 'low-quality content', including user accounts, posts, and weblinks selected for removal. The removal of social media posts and accounts thus constitutes the central line of action against influence operations, misinformation, false or fabricated news items, spam, and user-generated hyperpartisan news. While social platforms rarely disclose content that was flagged for removal, some companies have released publicly the community standards used to remove content from their services (Weedon et al, 2017; Facebook, 2018a, 2018b).

Studying the politics of deletion on social platforms is thus an exercise in back engineering, as content that has been deleted or blocked from social platforms is likely to be low-quality information. As such, the relative frequency of deleted social media posts and accounts can be used to gauge the extent to which election campaigns on social media were plagued by problematic content. The process of verifying if content remains available is, however, cumbersome. In addition to that, election campaigns need to be monitored in real time, as once a post is deleted by a user or blocked by the platform it disappears from the platform altogether; similarly, deleting a tweet automatically triggers a cascade of deletions for all retweets of that tweet (Twitter, 2018a). This specific affordance of social platforms facilitates the disappearance of posts, images, and weblinks from the public domain.

Ephemerality is, of course, a vital component of much social media communication, but it is not an expected or desirable design for political campaigns on social media. The disappearance of one-third of the discussion underpinning a key election campaign is particularly worrying, considering that the politics of deletion can be leveraged for targeted manipulation and disinformation. Indeed, much of the deleted content was caused by Twitter actively blocking the user accounts, thus suggesting that the Brexit debate may have been subjected to substantial amounts of low-quality information which is often a proxy for artificial manipulation and false amplification (Walker et al, 2019). The implications of observing such high decay in dynamic content is that it erodes the assumption that social media can serve as the public accord and public record of the decision-making process underpinning public deliberation (Papacharissi, 2008).

The large-scale removal of social media content has the problematic drawback of altering the record of social interactions. Unlike traditional media used to distribute propaganda, such as newspapers and posters in the 20th century (Sanders and Taylor, 1982), or pamphlets and leaflets going back as far as the 16th century (Raymond, 2003), the removal of social media content eliminates any trace of the event and ultimately prevents forensic analysis and academic research on influence operations targeting social media platforms. While there have been attempts to create public archives of social media posts, these institutional efforts faced considerable challenges and failed to come to fruition (Zimmer, 2015). Similarly, although social platforms have at times offered archives of disinformation campaigns identified and removed by the very platforms (Elections Integrity, 2018), such sanctioned archives offer only a partial glimpse into influence operations (Acker and Donovan, 2019).

Indeed, influence operations on social platforms, particularly on Facebook and Twitter, continue to rely on coordinated and targeted attacks where the accounts and profiles sourcing the content disappear in the months following the campaign. Some accounts are suspended by the social platforms for violating Community Standards and ToS (Gleicher, 2019), such as posting inappropriate content or displaying bot-like activity patterns; others are deleted by the malicious account holders, reportedly to cover their tracks (Owen, 2019). As discussed in Chapter Six, the modus operandi of influence operations often consists of amplifying original hyperpartisan content by large botnets that disappear after the campaign. The emerging thread is then picked up by high-profile partisan accounts that seed divisive rhetoric to larger networks of partisan users and automated accounts (Bastos and Mercea, 2019). These tactics suggest that tweet decay is a key metric to identify problematic content, including influence operation and false amplification on social media.

Influence operations can thus daisy chain multiple disinformation campaigns that are phased out and disappear as soon as rectifying information or alternative frames start to emerge. As previously discussed, the politics of deletion can be leveraged by propaganda campaigns centred around the firehose of falsehood model (Paul and Matthews, 2016). The high-volume posting of social media messages is effective because individuals are more likely to be persuaded if a story, however confusing, appears to have been reported repetitively and by multiple sources. Traditional counterpropaganda methods tend to be ineffective against this technique. Similarly, fact-checking social media posts that have disappeared is not technically possible and perhaps not desirable either (Vinhas and Bastos, 2022).

Orphaned data

Tweet and user decay, or the disappearance of tweets and user accounts from the platform, is a challenge for research on disinformation campaigns, as the content required for forensic analysis is likely to have disappeared from public scrutiny. For comparison, the 2019 UK General Election registered a relatively low rate of tweet decay. On the eve of the vote, only 6.7 per cent of election-related tweets had been removed and less than 2 per cent of the accounts were no longer operational. This figure is in line with previous studies reporting that on average 4 per cent of tweets disappear (Xu et al, 2013; Bagdouri and Oard, 2015), but contrasts with the Brexit database where we identified a much higher rate of tweet and user account deletion.

The process of verifying whether a social media post remains available after being posted online can be made via http requests or programmatically using a social platform's Application Programming Interface (API). Twitter allows developers to retrieve programmatically and at scale (that is, 'rehydrate') the full tweet, user profile, or direct message content using their APIs (Twitter, 2019a), but their ToS state that content deleted by a user or blocked by the platform due to infringements on the ToS ought to disappear from the platform. This specific affordance of social platforms has of course facilitated the disappearance of posts, images, and weblinks from the public view, with important and negative effects to research on influence operations.

Social platforms rarely disclose content that was flagged for removal, and therefore studying the politics of deletion on social platforms is an exercise in reverse engineering. Conversely, content that has been deleted or blocked from social platforms is likely to fall within the broad category of 'low-quality content', and therefore content decay – tweet and user decay in the context of Twitter – can be used as proxies to examine the extent to which deliberation on social media is hindered by influence operations, including disinformation and other forms of problematic content (Starbird et al, 2014, 2019).

There is surprisingly little research on how social media data sets change when observed at different points in time and how this may impact the results of the analysis. Walker (2015) contrasted data collected from social media platforms in real time versus data collected minutes, hours, or days after the post went online. McCreadie et al (2012) explored the effect of this collection decay on the Tweets2011 data, a set of 16 million tweets offered by Twitter and made available to participants of the TREC Conference. The data set, whose tweets cover the period of 23 January to 8 February 2011, was reconstructed using an asynchronous HTTP fetcher (as opposed to Twitter APIs) in 2012. Given the limitations and the unreliability of HTTP crawlers, no precise information was given regarding the ratio of tweets that disappeared within the one-year gap between the data being made available and the reconstruction of the data set.

Bagdouri and Oard (2015) also developed an evaluation design to predict whether a tweet will be deleted within the first 24 hours of being posted. The classifier, which takes into account the distribution of deleted tweets, along with tweet and user features, reported a very sharp skew in the data, with some users regularly deleting their tweets while others rarely did so. Bagdouri and Oard (2015) also found that the most prolific deleters were automated systems engaging in advertisement, a known marker of spambots that push low-quality content and are ultimately purged from social platforms. More interestingly, the authors reported a remarkably low ratio of tweet decay in the 24-hour period, with only 3.6 per cent of the messages having been deleted, or 2.2 per cent of the data once retweets were excluded. Finally, they reported that 2 per cent of users were responsible for just above one-third of all deletions that were not of retweets.

In related research, Xu et al (2013) collected a corpus of over 300,000 bullying-related tweets and estimated their survival rate (or persistence) by querying the URL of the tweet for around two months at regular intervals. The deletion rate found in the data, hypothesized to be a function of user's regret, allowed for the construction of a 'regrettable posts predictor'. Deletion rate was found to decay over time, with a drop-off in deletion rate that was so extreme that the authors could safely exclude deletions occurring after two weeks from the filtered data set without significantly introducing any noise. In the end, the overall fraction of deleted tweets was rather low and similar to those reported by Bagdouri and Oard (2015), at 3.75 per cent. This fraction of deleted tweets of 4 per cent is considerably lower than what we have anecdotally observed in a range of political and protest data sets. In fact, from our experience monitoring the Twitter Compliance Firehose, we estimate the baseline of tweet deletion to currently stand at around 15 per cent.

Tweet decay

Deletion rate in the Brexit database stands in sharp contrast to the figures reported in the literature. We inspected the period of 13 days leading up to the referendum vote on 23 June 2016, with a total of 2,775,789 messages tweeted by 792,663 users, thereby averaging around 3.5 messages per user. Social media activity presents a long-tailed distribution and Twitter is no exception to this: 169 users tweeted over 500 messages and 28 posted more than 1,000 messages. The campaign accounts @ivotestay and @ivoteleave posted 15,928 and 11,647 tweets in this 13-day period, respectively, and since the referendum both accounts have been suspended by Twitter. Indeed, we manually checked the 100 most active user accounts and 37 are no longer active: 16 have been suspended and 21 no longer exist (deleted account). From the remaining 63 active accounts, three were recreated after the referendum and two have been set to private. In short, only about half of the most active accounts in the referendum debate continue to operate publicly.

This trend is not restricted to hyperactive users. We queried Twitter REST API and the web interface to check which user accounts remained active in October 2019, three years after the vote. For the universe of 792,663 users that tweeted the referendum between 12 and 24 June, 20 per cent or 155,157 accounts are no longer active. These numbers are consistent whether checking by username or user ID, with only 118 mismatches (false positive or negative) when querying Twitter REST API by usernames instead of user ID. Twitter REST API does not inform if the accounts have been deleted, blocked, suspended, removed, or set to private – but this information can be retrieved from the web interface. We therefore checked the web interface of each of these accounts to identify accounts that have been suspended by Twitter. Twenty per cent of the accounts that are no longer active have been blocked by Twitter ($n=36,159$).

The figures are yet more dramatic when we inspect the share of messages that are no longer available on Twitter, whether on Enterprise, Search, or REST APIs, or via web interface. Twenty-two per cent, or 631,700, tweets are no longer available due to the removal of the seeding user and the enforcement of the ToS governing content authored by deleted accounts. Posts from deleted accounts are retrospectively removed from Twitter and generate orphaned data, but this number only accounts for messages that are no longer available due to the user account that posted the content no longer being active. We rehydrated the data to account for this difference, thereby identifying messages that were actively deleted by users, as well as retweet cascades that disappeared due to the original seeding post no longer existing in the platform.

One-third of the near 3 million tweets posted in the period are no longer available ($n=932,815$), with only 1,842,974 tweets remaining. In other words, 33 per cent of the tweets that shaped the discussion about the referendum are no longer retrievable three years after the vote and nearly half of this universe of messages disappeared because the seeding account was removed, blocked, deleted, suspended, or set to private. One-fifth of the tweets that are no longer available disappeared because the seeding account was suspended from Twitter. This figure ($n=203,681$) includes not only tweets authored by these accounts, but also intermediary points in retweet cascades that vanished because the user account was no longer available. Suspended accounts were particularly prolific: while less than 5 per cent of users were suspended, these accounts posted nearly 10 per cent of the entire conversation about the referendum on Twitter.

Figure 9.1 unpacks the fraction of deleted tweets and accounts: around 20 per cent of users (or 155,157 out of 792,663) are no longer active and, again, 20 per cent of this cohort of accounts have been blocked by Twitter (or 36,159 out of 155,157). Twitter API, unfortunately, offers no other information about the remaining 118,998 accounts that are no longer active. These may be accounts that have been compromised, employed in influence operations, or that violated Twitter's ToS, but unfortunately only the information available in the API can be made public due to the constraints of their Privacy Policy (Twitter, 2019b). This is a considerable limitation, as this small cohort of accounts posted nearly 10 per cent of the entire conversation about the referendum on Twitter. As such, it is conceivable that this small cohort of removed accounts may have sourced low-quality content at scale and speed.

We also found that both #voteleave and #voteremain account for a significant share of the hashtags in the data, with the former appearing in 15 per cent of the tweets and the latter in just under 10 per cent. The hashtag #voteleave is, however, significantly more likely to appear in tweets that are no longer available. Indeed, 20 per cent of the deleted messages espoused this term compared with the regular 10 per cent for #voteremain. No other hashtag presents such an enormous difference between the frequency observed in the population of deleted tweets compared with the frequency observed in the entire population of messages. The point difference observed for #voteleave is comparable to the point difference of the remaining 700 most popular hashtags combined ($x=.064$ for both groups).

Figure 9.1: Tweet and user account decay

(a)

Tweets posted during the referendum campaign June 12–24

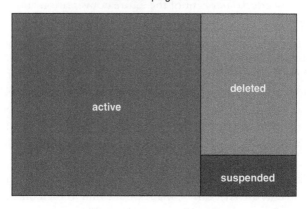

(b)

Twitter accounts that tweeted the referendum June 12–24

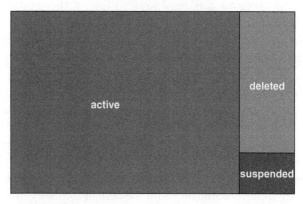

Note: (a) 33 per cent of the tweets leading up to the referendum vote are no longer retrievable; (b) 20 per cent of the user accounts that tweeted the referendum are no longer active.

One important caveat is that the Vote Leave campaign accounts for a larger share of the tweets due to the prevalence of popular Leave hashtags. We, therefore, controlled for the higher number of tweets and users associated with Leave, but Leave users continued to be identified as more likely to be deleted, blocked, or removed from Twitter. Similarly, the tweets posted by these accounts are considerably more likely to decay. Nearly 40 per cent of the messages posted by Leavers are no longer available, whereas the fraction of deleted tweets for Remainers is under 30 per cent. If we assign the hashtag #brexit to the Leave campaign, we find that more messages from the Leave campaign disappeared from Twitter than made up the entire universe of messages affiliated with the Remain campaign: 468,419 tweets disappeared from a total of 1,224,568 messages posted by Leavers, and 130,245 posts are no longer available from a universe of 438,359 messages posted by Remainers.

As discussed in Chapter Four, the overall sentiment during most of the campaign was decidedly nationalistic, with three-quarters of messages having a nationalistic sentiment. Most messages tweeted in the period were additionally preoccupied with economic implications of the decision to leave the EU. There are however significant differences in the ideological orientation of the Brexit debate when controlling for users and tweets that have since disappeared. While the debate is predominantly focused on economic issues, most of the messages that are no longer available (n= 932,815) largely appeal to populist slogans ($\bar{x}$ = .52 and $\tilde{x}$ = .50) compared with the subset of messages that remain active ($\bar{x}$ = .44 and $\tilde{x}$ = .43). This substantial difference is again observed in the ideological polarity Globalism vs Nationalism. The debate is indeed predominantly nationalistic, but it is significantly more so in the subset of messages that are no longer available: $\bar{x}$ = .64 and $\tilde{x}$ = .67, compared with $\bar{x}$ = .59 and $\tilde{x}$ = .63 for the subset of messages that remained active.

We compared these figures with a control group of tweets collected in the same period of early 2016. To this end, we rehydrated a range of hashtags with several thousand tweets posted in the period of the official campaign of the UK EU membership referendum. For each hashtag, we take a random sample of 1,000 tweets in the data set and query Twitter API to verify if the message is still available and whether the account (by user ID) that sourced the content remains active. A total of 45 hashtags were queried in five groups of nine hashtags for the Remain[1] and the Leave[2] campaigns, nine non-partisan hashtags discussing the referendum,[3] nine non-political hashtags[4] that trended in the period, and nine hashtags dedicated to protest activism[5] causes that also trended at the time (Figure 9.1).

We found that 15% was the baseline for deleted tweets in 2016 (Q_1=.10, $\bar{x}$ = .15, $\tilde{x}$ = .15, Q_3=.19) based on the ratio of deleted tweets in Group 4, which includes a set of non-political hashtags. This baseline increase sharply as the topic of the conversation becomes more contentious. Protest activism hashtags present a tweet decay of nearly 30% for #15maydebout and 44% for #blacklivesmatter, a hashtag campaign notable for being targeted by the Russian IRA operations (Bastos and Farkas, 2019; Freelon et al, 2020). Brexit related discussions also verge around 30% (Q_1=.24, $\bar{x}$ = .28, $\tilde{x}$ =.29, Q_3 = .32), a figure that is not too different from what was observed in openly partisan hashtags associated with the Remain campaign (Q_1 = .23, $\bar{x}$ = .30, $\tilde{x}$ = .26, Q_3 = .32). Openly partisan hashtags associated with the Leave campaign, however, present

---

[1] Group 1: *betteroffin, bremain, leadnotleave, lovenotleave, moreincommon, strongerin, votein, voteremain, yes2eu.*

[2] Group 2: *beleave, betteroffout, britainout, leaveeu, loveeuropeleaveeu, no2eu, notoeu, voteleave, voteout.*

[3] Group 3: *brexit, brexitornot, antibrexit, euref, eureform, eurefresults, jocoxrip, lovelikejo, referendum.*

[4] Group 4: *mozfest, nsmnss, rstats, agchat, cadrought, foodsystem, homegrown, phdchat, usaid.*

[5] Group 5: *15maydebout, 7mdebout, bruxellesdebout, globaldebout, nobillnobreak, nuitdebout, romadebout, blacklivesmatter, lesvos.*

a worrying and much higher coefficient of tweet decay ($Q_1 = .33$, $\bar{x} = .42$, $\tilde{x} = .42$, $Q_3 = .50$).

Indeed, three-quarters of the content hashtagged with #betteroffout has been removed from Twitter. More than half of the tweets hashtagged with #voteleave, the official campaign to leave the EU, is no longer available. Figure 9.2 unpacks the differences across classes of hashtags. The size of the line indicates the discrepancy between account deletion rate and tweet deletion rate. In other words, the longer the line, the higher the fraction of deleted tweets relative to the fraction of deleted accounts. This can be caused by users regretting and deleting the original post, by having their accounts repurposed, or else by users changing their account to private, in which case the original tweet is no longer available even though the account remains operational.

We validate this analysis by parsing the hashtags featured in the pre-Brexit data set and calculate the deletion rate per tag. Tweets hashtagged with terms associated with the Leave campaign are considerably more likely to have been deleted. In fact, the list of hashtags tweeted over 1,000 times with a deletion rate of 40 per cent or higher includes mostly Leave hashtags, with only a few terms not clearly aligned with either side of the campaign: *voteleave, votein, leaveeu, ivoted, voteout, beleave, cameron, inorout, ukip,* and *eng*. For this set of hashtags, most of the messages tweeted in the period leading up to the vote are no longer available ($\bar{x} = 52\%$).

Ephemeral hyperpartisanship

Upon inspecting the Brexit database we found that nearly 3 million messages posted by 1 million users in the last days of the Brexit referendum campaign were no longer available (Bastos, 2021c). While previous studies have found that on average only 4 per cent of tweets disappear (Xu et al, 2013; Bagdouri and Oard, 2015), we found that 33 per cent of the tweets leading up to the referendum vote have been removed.

Figure 9.2: Percentage of deleted user accounts and tweets

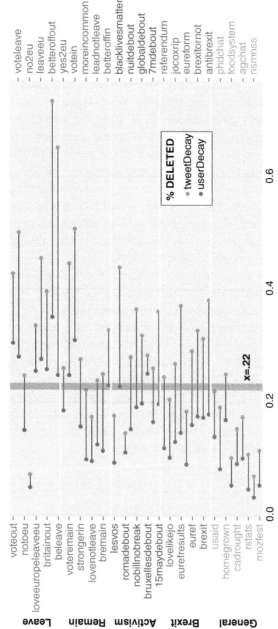

Note: Dark dots – fraction of deleted accounts (percentage points); light dots – fraction of deleted tweets (percentage points). The tweet deletion rate for hashtags betteroffout and beleave are 74 per cent and 65 per cent, respectively. Vertical grey bar shows the mean deletion rate for this set of 45 hashtags.

Only about half of the most active accounts that tweeted the referendum continue to operate publicly and 20 per cent of all accounts are no longer active. These accounts were particularly prolific: while Twitter suspended fewer than 5 per cent of all accounts, these posted nearly 10 per cent of the entire conversation about the referendum.

We also tested whether partisan content was more likely to disappear and found more messages from the Leave campaign that disappeared than the entire universe of tweets affiliated with the Remain campaign (Bastos, 2021c). In fact, the list of hashtags tweeted over 1,000 times with a deletion rate of 40 per cent or higher is largely restricted to Leave terms. For this set of hashtags, most of the messages tweeted in the period leading up to the vote are no longer available ($\bar{x}$ = 52%).

Tweet decay also appears to be dependent on the timeframe of the Brexit negotiations and deliberation. To this end, we queried a database of 100 million Brexit-related tweets posted by users based in Britain (British Twitter Monthly Active Userbase – BTMAU). The database encompasses the official campaign period for the 2016 referendum campaign until October 2019 and includes 43 months of Brexit-related messages. We took a sample without replacement of 50,000 tweets per month, therefore generating a data set of 2,150,000 tweets. We proceeded by querying Twitter REST API to check whether users and tweets were still active in the platform. This approach allowed us to calculate a coefficient for tweet decay that is representative for each of the 43 months that followed the referendum campaign.

In the weeks leading up to the vote, deletion rose from 19 per cent to 33 per cent. Tweet decay receded after the referendum and only resumed when pressure started to mount for triggering Article 50, at which point the monthly fraction of deleted tweets peaked from 27 per cent to one-third. After Article 50 was triggered, the fraction of deleted tweets stabilized and only escalated again when then-Prime-Minister Theresa May announced the Chequers Plan, a contentious proposal whose deliberation took the fraction of deleted tweets to around one-fifth. Tweet decay decreased in the ensuing months but only became similar to the figures reported in previous studies in the premiership of Boris Johnson, when tweet decay was the lowest at 7 per cent. Figure 9.3 illustrates these trends.

However, tweet decay was rarely as low as 4 per cent, the maximum estimate found in studies prior to 2016. From dozens of hashtags archived in 2016 and rehydrated in 2019 (Twitter, 2019a), we established that 15 per cent was the deletion baseline for non-political, uncontroversial hashtags, a baseline that increases sharply as the issue at stake becomes contentious. Tweet deletion in protest hashtags tweeted worldwide was similar to non-partisan Brexit hashtags at nearly 30 per cent. Tweet decay in openly partisan hashtags associated with the Leave campaign, however, was much higher, at 42 per cent on average.

Indeed, three-quarters of the content hashtagged with #betteroffout has been removed from Twitter and more than half of the tweets hashtagged with #voteleave, the official campaign to leave the EU, are no longer available. While the volume of political content removed from Twitter is astonishingly high, there is also evidence that social platforms are removing more content in general and systematically purging accounts. In sharp contrast to the 4 per cent baseline of tweet decay reported before 2016, we found that even non-political messages are being purged at a rate at least three times as high. As such, it is difficult to rely on social media as a record for public deliberation, as the public record disappears with no recourse for recovery.

These anomalies are promptly detected by the Seasonal Hybrid ESD (S-H-ESD) algorithm (Vallis et al, 2014). Anomalies are found in the series of active and deleted tweets on the date of the referendum and the triggering of Article 50, which is unsurprising given the increase of activity reflecting these key dates in the Brexit calendar.

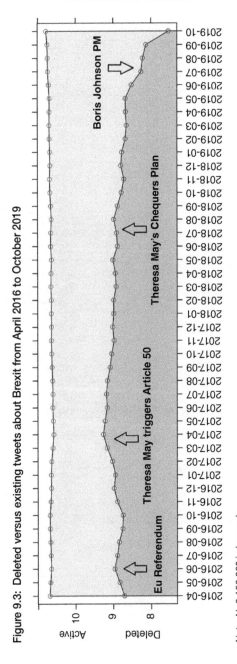

Figure 9.3: Deleted versus existing tweets about Brexit from April 2016 to October 2019

Note: N=2,150,000 in log scale

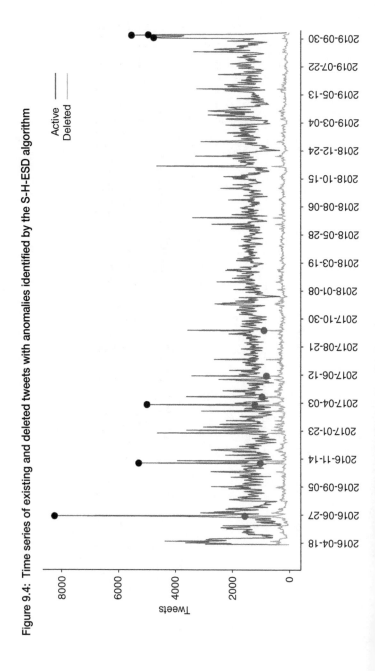

Figure 9.4: Time series of existing and deleted tweets with anomalies identified by the S-H-ESD algorithm

There are, nevertheless, significant global anomalies in the series of deleted tweets that are not present in the series of active tweets. These can be pinpointed to the large volumes of tweets that disappeared after key dates. Indeed, a substantial number of messages disappeared three weeks after Article 50 was triggered, and again another significant upsurge in tweet deletion was registered the day after the snap general election (8 June 2017). None of these anomalies are registered in the regular series of tweets that remained in the series, which is consistent with the assumption that tweet decay is associated with politically contentious contexts. In fact, most of the anomalies in the series of existing tweets were registered in late 2019, when Boris Johnson became prime minister and parliamentary elections were called for 12 December 2019. Figure 9.4 illustrates these differences, showing the temporal series of existing and deleted tweets and the anomalies detected in these series.

TEN

Accountability of Social Platforms

This closing chapter discusses the public accountability of social media platforms that have eluded policy makers. We discuss the challenges in detection and mitigation of misinformation that emerged in the Brexit debate on social media. We also discuss the growing perception that social networking sites are opaque platforms unaccountable to regular users and governments alike, with sentiments towards Twitter bots and trolls often taking centre-stage. We argue that social media algorithms, in particular Facebook and YouTube recommendation engines, remain largely unaccountable to public scrutiny and that the criteria underpinning algorithmic decisions on which news stories are distributed to users are intellectual property deemed commercially sensitive and therefore inaccessible to the public.

The right to remember

The data we analysed in the previous chapter caution against the use of social platforms as a complement or extension to deliberative forums, as the public record of social media interactions, or at least considerable portions of it, is subject to being removed from public scrutiny. This is particularly problematic if one assumes social platforms may offer a substitute for deliberative forums of civic participation. The theoretical import of the Habermasian perspective to rational public deliberation assumes that the collective decision-making on issues of public concern resolves conflicts based on the quality of the argument and the evidence supporting it (Vinhas and Bastos, 2022). Our analyses of the Brexit referendum, at least as far as the discussion was registered on Twitter, are largely at odds with these stipulations. While the volume of Twitter activity may be directly linked to developments unfolding over time (Bastos et al, 2015), we found that a significant portion of the Brexit debate on Twitter was not designed for permanence.

Ephemerality is, perhaps, an expected affordance of social media communication, but it is not an expected design of political communication and deliberation across social platforms. The disappearance of one-third of the discussion underpinning the Brexit referendum indicates that the fraction of deleted tweets may be a proxy for manipulation and disinformation. As much of the deleted content resulted from Twitter actively blocking user accounts, and in doing so generating orphaned data, it is conceivable that the Brexit debate may have been subjected to considerable volumes of low-quality information whose distribution often resorts to artificial manipulation and false amplification (Walker et al, 2019). In other words, while the ephemerality of social media posts may be a reasonable expectation of users, this poses considerable challenges for informed public deliberation around matters where the issue being deliberated on is constantly disappearing from public scrutiny.

Unfortunately, the identification of removed content and user accounts entails computational routines that cannot be implemented in real time, as there are multiple triggers that may block or delete an account and post from social platforms. Influence operation may conceivably exploit these limitations by offloading problematic content that is removed from platforms before the relentless – though time-consuming – news cycle has successfully corrected the narratives championed by highly volatile social media content. In Chapter Eight we described this process as the involuntary but spontaneous gaslighting of social platforms: the low persistence and high ephemerality of social media posts are leveraged to transition from one contentious and unverified political frame to the next before mechanisms for checking and correcting false information are in place.

In other words, while social media content may be fundamentally ephemeral, the fraction of deleted tweets we identified in the Brexit database was disturbingly high and prevented further analysis of the accounts that seeded the content, as once users are removed from the platform no further information can be gleaned from the account. The implications of observing such high decay in dynamic content at the centre of political discussion certainly warrants further consideration from policy makers. For academic researchers, it would be important to know the cut-off point after which political content is likely to disappear from social platforms. While previous studies show that decay is associated with time, data from the Brexit debate show that political and especially contentious messages are more likely to disappear from the public record than non-partisan conversations recorded in the same period.

Further research is also required to establish the extent to which different forms of political discussion are equally likely to disappear from social platforms, which temporal patterns are indicative of survival, and whether decay is caused by manipulation and influence operations detected by and ultimately removed from social media by the platforms themselves. This may require research collaboration with social media platforms, especially if the objective is to establish whether the decay in social media posts is associated with influence operations and low-quality content, which reportedly is characterized by a shorter shelf life compared with organic content.

Another important empirical question that could not be addressed in the course of this book is determining the exact point in time when accounts and tweets sourcing political content are likely to start decaying. While Xu et al (2013) have found a strong and inverse linear association between the fraction of deleted tweets and time (in one-minute increments), this may apply only to the corpus of bullying tweets studied by the authors. The extreme drop-off in deletion rate reported by the authors, along with the remarkable low ratio of deleted tweets at only 4 per cent, are important indicators that social media decay likely differs across topics and may be substantively higher for conversations targeted by influence operations. In other words, one important empirical question that warrants further research is whether deletion rate is universally and inversely associated with time.

Problematic and elusive content

Much of the public debate revolving around social media manipulation relies on the contested and ideologically inflected notion of 'fake news' (Farkas and Schou, 2018). While scholarship often circumvents these problems by employing instead the term 'false news' (Vosoughi et al, 2018), the vernacular use of 'fake news' has the advantage of encompassing a broad spectrum of content, including influence operations, mis- and disinformation, but also false amplification, coordinated inauthentic behaviour, and the use of botnets to manipulate public opinion. Social platforms, therefore, rely on the broad term 'problematic content' to refer to a wealth of content that takes advantage of the biases (Comor, 2001; Innis, 2008) intrinsic to the business model of social platforms based on the steady supply of viral content.

Problematic content and the intentionally broad term 'low-quality information' thus incorporate misinformation (and subcategories such as disinformation), influence operations that mimic the appearance of news outlets, false or fabricated news items, and user-generated hyperpartisan news – that is, polarized narratives reinforcing partisan identity. Hyperpartisan content is particularly difficult to classify because the intentions of the sender, the key category separating misinformation from disinformation, are at best elusive on social media. In addition to that, the style, messaging, lifespan, and provenance of low-quality information and hyperpartisan content can vary depending on the strategies employed to sow division. Social media platforms addressed these challenges by implementing a range of measures to identify false amplification (Weedon et al, 2017) and remove 'fake accounts' seeding problematic content (Twitter, 2018b), an operational term employed to describe accounts, posts, and links to content selected for removal. They have similarly implemented community standards that define problematic content and the enforcement of such standards (Weedon et al, 2017; Facebook, 2018a, 2018b).

Low-quality information in the Brexit database includes textual or audiovisual content tailored to nudge or interfere with online deliberation by deceiving, confusing, or misinforming. This definition emerged through our systematic analysis of the text corpora and the images distributed by accounts identified as Brexit Bots. This definition has the added benefit of being agnostic to the veracity or falsehood of statements, drawing instead from the field of risk innovation to identify threats to values within social and organizational contexts (Maynard, 2015). While social platforms define problematic content as information more likely to be removed, our approach identified information as low-quality whenever a repertoire of tactics was employed to hinder online deliberation (Halpern and Gibbs, 2013).

Indeed, the tweets in the Brexit database often included divisive content that blurred the lines among misinformation, disinformation, and propaganda. These were often user-generated content, but presented with tabloid-like formatting that included the expert use of sensationalized language and speculative reasoning (Bastos, 2016). Another marker of this type of content in the Brexit database is that they are likely to be eventually deleted or blocked from Twitter, only to resurface through sockpuppet or surrogate accounts that repost the content by linking it to a different webpage.

Webpages linked to Brexit tweets were particularly susceptible to decay, disappearing or significantly changing shortly after being posted, with a significant share of the webpages posted in the course of the referendum campaign disappearing shortly after the ballot (Bastos and Mercea, 2019). The webpages that could not be resolved after the referendum either linked to a Twitter account that had been removed, blocked, or deleted, or to a webpage that no longer existed. Nearly one-third (29 per cent) of the webpages posted before the referendum linked to multimedia content, such as Twitter statuses and pictures, that was no longer available and whose original posting account had also been deleted.

The low-quality information that circulated during the Brexit debate was marked by a remarkably short shelf life, which makes it difficult to retrospectively rebuild the public conversation. This problem was compounded by the extensive use of images on social platforms, which are associated with information cascades (Dow et al, 2013; Cheng et al, 2014), and the Terms of Service requiring content deleted by a user to be removed from social platforms (Twitter, 2018a). This combination of social media affordances and their corporate policies enables the disappearance of messages and linked content from the public domain and severely hinders research on mis- and disinformation campaigns. Public web archives, such as the Internet Archive, rarely contain records of specific tweets, their attached images, and linked content. As a result, it is often impossible to determine what the original social media post and associated image conveyed at the time of posting, or how far afield it was shared.

Repertoires of problematic content

The results of the Brexit project reported in this book foreground a range of metrics, indicators, and signatures that social platforms are likely to have implemented to maximize the early detection of problematic content. Our work on the Brexit database indicates that low-quality information shares a repertoire of features that allow for classifying it at scale and speed. While no single similarity measure or training data set can account for the variety of approaches available to mis- and disinformation outlets, such campaigns are likely to source content of quantifiably lower quality compared with mainstream media articles, which are labour intensive and reviewed by editors and therefore more expensive to produce.

Instead of metrics, it would be more accurate to describe the repertoire of problematic content relative to the absence of common traits found in high-quality content. These include stylistic and temporal signatures resulting from the reduced editorial practices and journalism strictures. The implementation of large language models (LLMs) may conceivably identify problematic content based on training data of hyperpartisan content, which could then be parametrized to systematically draw distinctions based on stylistic similarities of documents based on terms, term combinations, sentence length, punctuation, or parts of speech. We also found that problematic and low-quality content distributed in the Brexit campaign was produced in bulk for quick turnover and characterized by a short shelf life, an important point of departure from quality content that requires extensive editorial work at various levels of the news industry and is designed for durability (Bastos et al, 2021b).

The incidence of ephemeral and partisan content is not restricted to contentious issues such as the 2016 US presidential election or the 2016 UK EU membership referendum. But we found in the course of our research that, at least for this subset of the data, content that tends to disappear from Twitter is largely hosted by social media platforms and content curation services. Other than twitter.com, the most common domains include bit.ly, youtu.be, goo.gl, instagram.com, facebook.com, breitbart.com, and paper.li, with nearly half of the content published using paper.li disappearing shortly after it was made public. Paper.li is only one of many services that generate a 'professional looking newspaper' from user-generated content, another marker of low-quality content as misinformation sources often model their website layouts after established news outlets.

In addition to its ephemeral dimension, with many tweets and webpages posted in the Brexit saga having since disappeared, this content is often user-generated and therefore less likely to originate from authoritative sources of content. Other markers of low-quality content include stylistic devices and vague statements common to daily communication, which represent a considerable departure from curated content sourced from mainstream news outlets characterized by editorial and curation processes. Low-quality tweets that would eventually disappear often featured 'hashtag stuffing' – that is, they included in the text body several hashtags as a marker of emphatic partisan loyalty. The tweets often lacked sentence structure and contained many nouns but few adjectives as well as a great deal of conditional language that deviates from information that has been edited, curated, or reviewed for quality, accuracy, and persistence.

Low-quality information is, therefore, less diverse compared with quality content which is marked by accuracy, grammar and spelling consistency, and a richer vocabulary (Lijffijt et al, 2016). These are known features of low quality information marked by less-well-edited text, highly partisan discourse, and user-generated content that is quickly modified or erased; measures that are likely to identify not only misinformation but also the broader universe of problematic content. Despite these features and rhetorical devices that are typical of low-quality content, indeed despite Twitter taking down these tweets, the content would routinely remain available on Twitter even after the original seeding account was blocked or the webpage sourcing the content was removed. This is because deleted content resurfaces through new accounts that repost the original link to other similarly hyperpartisan websites, with deleted webpages generating several sister webpages, so that low-quality information continues to thrive on social platforms through reposting and resharing (Bastos et al, 2021b).

Last, the hyperpartisan dimension of deleted Brexit tweets was typically associated with bias or ideological alignment, resorting to rhetorical features that express certainty, speculation, and stylistic qualities of text data, such as the mood of verbs and use of conditional phrases or future tense. This partisan dimension could, therefore, be identified based on linguistic signals employed in the compositions, such as the absence of subjunctive mood to express uncertainty (Jiang and Wilson, 2018), specific rhetorical indicators of certainty (Kuklinski et al, 2000), and affective validation (Rucker et al, 2014). The predominance of nouns and the paucity of adjectives foreground less-nuanced content expressing partisan certainty, whereas quality information contrasts established facts and conditional possibilities, a metric that has long been leveraged to identify common expressions in hate speech, false news, and texts associated with sex, death, and anxiety, as opposed to words related to work, business, and the economy (Pérez-Rosas et al, 2017).

Algorithmic junk

Central to the distribution of polarizing, hyperpartisan Brexit content was the role played by algorithmic systems that underpin Twitter, but that of course extend to the Facebook News Feed and YouTube suggested videos. As digital platforms such as Twitter become critical intermediaries in the distribution of news (Bastos, 2015), their recommender systems can stimulate or inhibit specific subsets of news content that remains essential to the functioning of democratic institutions (Schudson, 1995). Indeed, the rationale underpinning these algorithms stems from the priorities of the social platform business model, centred on advertisement, rather than the strictures of journalism practice. These recommender algorithms filter, rank, and select which content is shown to users (Resnick and Varian, 1997) through an infinite scrolling of individual pieces of content that include news items, status updates, photos, web links, and videos.

Distilling many hundreds of pieces of possible content, the algorithmic distribution of news selects content that is optimized for user engagement, which often translates to prioritizing content that taps into primal emotions, such as anger or fear (Ananny and Crawford, 2018). The disclosures of Facebook whistleblower Frances Haugen show that the news industry raised concerns about changes made to the News Feed directly with Facebook back in 2018, when Jonah Peretti, CEO of Buzzfeed, raised the issue of a significant algorithmic change in favour of 'meaningful social interactions' that translated to 'pressure to make bad content or underperform', as the recommender system favoured 'fad/junky science' and 'extremely disturbing news' (Hagey and Horwitz, 2021). Within this closed environment of algorithms that are agnostic to hatred and vitriol, reality-distorting misinformation can flourish on social platforms by consciously and reliably tapping into users' darkest impulses and polarized politics (Bastos, 2022a).

Algorithmic changes made to Twitter's Trending Topics and other recommender systems are likely to have had profound effects on the gathering, composition, and distribution of information tailored for the Brexit campaign (Thurman et al, 2019). In choosing what content on offer will be presented to the end user, the algorithms take on critical roles of gatekeeping and agenda-setting functions traditionally the domain of trained journalists in the news media industry (DeVito, 2017). Twitter, but also Facebook and Google, have therefore become de facto media companies with gatekeeping power and agenda-setting function (Moore, 2016; Napoli and Caplan, 2017; Flew, 2019).

The tendency of social media users to show a preference for a subset of content that is at odds with the coverage of newspapers was already apparent before social media became the prevailing media for news consumption (Bastos, 2015), but the social algorithmization of news certainly worsened its effects. Benkler et al (2018) argued that it was Facebook algorithms – more than Facebook communities or specific malicious actors distributing problematic content – that rewarded clickbait websites and tabloid-like sources of information, which often include hyperpartisan content. The algorithmization of social media communities was particularly damaging because it reinforced patterns of interaction and the sharing of content in tightly clustered communities that supported and likely reinforced the relative insularity of users.

The conditions for online communities to trigger dynamics that deviate from those observed in face-to-face communities are likely to grow with the algorithmization of community governance implemented by social platforms. Smartphones and internet-connected devices continuously store individual trace data in the cloud. These data are imperfectly processed by machine-learning algorithms of limited precision and recall, trained with incomplete data sets, and tailored for purposes that are indistinguishable from those of the company collecting the data. By aggregating digital trace data at individual and group levels, social platforms can offer advertisers granular targets benchmarked with incomplete data and limited precision. These imperfections cause misclassifications and malfunctions largely understood as acceptable downsides of the social media business because imprecisions measured at the individual level are offset at the group level. But these imprecisions and incompleteness are then built into the system recursively, so that social groups identified by and modelled with digital trace data can take up a life of their own on social platforms, with niche subcommunities that only partially meet their physical counterparts (Bastos, 2022b).

The rise of large social media platforms such as Twitter has, therefore, followed the decline of the news industry in terms of revenue and profitability. The asymmetric power exerted by social platforms on news organizations is such that news companies find themselves trying to meet the ever-changing demands of social platforms' algorithms. These issues have profound implications for the public discourse and the informed citizenry that sit at the centre of democratic deliberation, an arena now populated by digital intermediaries such as Facebook, Google, and Twitter whose algorithms are unwitting lynchpins defining digital policy and the integrity of information online (McNally and Bastos, 2023).

ELEVEN

Conclusion

This concluding chapter takes stock of the Brexit referendum as a milestone in the political realignment that moved Western democracies towards nationalistic and populist sentiments, a development accompanied by the upsurge in affective polarization, proattitudinal partisan news consumption, echo-chamber communication, and coordinated and inauthentic social media activity. We argue that the Brexit debate on social media provides a bird's-eye view to political developments that would define much of the political landscape in the early 21st century and that are likely to continue defining much of the contentious politics in the first decades of this century, as technological change continues to trigger cultural, economic, and legal disruption.

In Chapter Two, we unpacked the British Twitter Monthly Active Userbase (BTMAU), a database counting 3.6 million UK-based users who posted content about Brexit. This cohort is more likely to be younger, urban, and politically engaged, but also white, well-educated, and wealthier. By triangulating information from their geocoded tweets, geolocation information from their user profiles, and information that appeared in their tweets, we established that the geographic distribution of the Brexit tweets across the UK is, unsurprisingly, concentrated in England, with London accounting for 19 per cent of the BTMAU, followed by Manchester, Glasgow, Bristol, Edinburgh, Leeds, Birmingham, and Liverpool. We also described the training and implementation of machine learning algorithms to identify the ideological value space of

Brexit using a model informed by the hypotheses advanced by Inglehart and Norris (2016) of economic insecurity and cultural backlash, the steps taken to identify echo chambers as a function of homogeneous partisan interaction, and a bot-detection protocol that identified suspicious automated activity.

The great upset

The election upset that marked the end of Britain's membership of the EU highlighted the ideological fault lines of an electorate divided across economic and cultural issues, with the Conservative Party embracing a nationalistic rhetoric and the Labour Party consolidating its grip on the liberal and metropolitan elite. This reorganization of the British political space led to a cleavage between the young and well-educated, who embrace progressive values, and the older, less-educated portions of the population who perceived a decline in their material conditions. It was against this backdrop of polarization and ideological incongruity that political campaigning on social media took centre-stage, either as a data source for modelling the results of the vote, or as a strategic tool for microtargeting undecided voters.

The geographic patterning of the vote to leave the EU reflected these socioeconomic imbalances between an affluent metropolitan elite clustered around London, and economically worse off parts of England and Wales. This was compounded by the nationalist cleavage between the British Government and devolved governments in Scotland, Wales, and Northern Ireland. Social media, Facebook and Twitter in particular, proved important in driving the UK vote to leave the EU by riding a wave of cultural backlash by older, traditional, socioeconomically deprived and less-educated voters, or by amplifying a broad sense of economic insecurity experienced by blue-collar working-class voters. The hyperpartisan content distributed by Brexit-leaning accounts on Twitter voiced this disenchantment towards elites while reasserting ethnic values and the role of the state against global economic integration, a set of values embodied in the 'Take Back Control' slogan of the Leave.EU campaign.

Our analysis of the Brexit tweets foregrounds the role of nationalistic sentiments, but also the economic cleavage that was integral to the Brexit rhetoric. Tweets from users based in constituencies with overwhelming support for the Leave vote presented predominantly nationalist sentiments, compared with 17 of such constituencies that displayed a Twitter debate predominantly defined by populist sentiments. Our classifier also found economic issues, and the policy implications of the vote to leave the EU, to have motivated the userbase tweeting the referendum, a component of the Brexit debate overshadowed by the much-discussed cleavage between the metropolitan elite in London and parts of England and Wales that were economically worse off.

Bots talking to bots

Bots are automatic posting protocols that can be used to distribute content at scale and speed. While Twitter is a relatively bot-friendly platform and most automated accounts are benign, bots can be leveraged to impersonate a third party and are associated with sockpuppets and false online identities used to manipulate public opinion. The literature investigating bot activity is concerned with the imitation of human activity, as the successful implementation of bots can approximate human conduct and influence communication exchanges on polarizing topics. This body of work also presents compelling evidence that political bots produce more positive content in support of a candidate. Identifying Twitter bots is, however, a challenging endeavour, even if our approach to bot detection was ultimately successful with Twitter having confirmed they removed 71 per cent of the accounts we identified as bots.

Concerns about the use of bots and sockpuppets in the Brexit referendum were expressed in the press and academia.

CONCLUSION

Howard and Kollanyi (2016) argued that the EU referendum bots were designed to take sides in the debate about the UK's membership of the EU. These concerns were often conflated with the disinformation campaign led by the Kremlin-linked and St Petersburg-based Internet Research Agency (IRA), an operation that relied primarily on supervised accounts deployed on Twitter, Facebook, and Instagram. By all accounts, the IRA seems to have been successful in galvanizing partisan communication online and agitating for rallies across the US, often contacting campaign staff members in various US states and appealing to individuals to take their grievances to the streets.

The Brexit Botnet we found in our database exploited the fertile ground of political realignment revealed by the referendum. It comprised a network of 13,500 Twitter accounts that tweeted extensively about the Brexit referendum, with a clear slant towards the Leave campaign. The sophistication of the operation identified in the botnet deviates from traditional Twitterbots, with curated replication of content that was both user-generated and an approximation of tabloid journalism. The overall tone of the messages was much in line with the context of disaffection with immigration and the cultural backlash spearheaded by the older, traditional, and less-educated readership of tabloids. This cultural backlash was strategically exploited with populist frames to promote traditional cultural values, particularly those with nationalistic and xenophobic undertones.

The Brexit Bots exhibited patterns of specialization that allowed them to trigger small to medium-sized cascades in a fraction of the time required by active users. Their influence in the network was, however, limited and their content was marked by a short shelf life and a form of storytelling that blurs the line between traditional tabloid journalism and user-generated content, often presented as a professionally looking newspaper by resorting to content curation services. The Brexit Bots were, nonetheless, effective at creating and joining Twitter cascades compared with regular Twitter users. Unsurprisingly, the Brexit Bots could trigger cascades much faster than real users, but the upper threshold of their cascades was still somewhat limited by human interaction, as they often retweeted content posted by real users as a strategy to maximize exposure to organic content.

Public concerns about the weaponization of social media for political campaigning stem not only from the perception that these platforms are manipulated by bots and trolls. These concerns are also rooted in the perception that the affordances of social platforms reinforce ideologically homogeneous interaction, whether with users one is more likely to agree with (echo chamber), or with algorithms that select and filter content based on one's partisan leanings (filter bubbles). As such, the echo chamber hypothesis posits that users on social media are more likely to interact with politically aligned others, and to share content that resonates with their ideological orientation. While the evidence supporting filter bubbles on social media is inconclusive, there is a substantive body of evidence based on digital trace data supporting the hypothesis that echo chambers are pervasive on social media, as social interaction is more likely to be driven by ideologically congruent social environments compared with the baseline of social interaction offline.

But the evidence for echo chambers reported in the literature is also mixed, with studies that relied on self-reported data finding no such effect, in contrast to studies relying on digital trace data where the prevalence of echo-chamber communication is well documented (Terren and Borge, 2021). Our hypothesis, conversely, was that echo chambers could be driven by group formations inherited from offline social interaction as opposed to being rendered from online activity alone. With echo chambers referring to closed communication systems where users interact, it is possible that users meet online to continue conversations they experienced offline, or to establish a pattern of interaction that resemble their day-to-day routines. It is also possible, of course, that these interactions were distorted by the large-scale deployment of social bots, whereby botmasters would manage a large group of Twitter accounts to tweet and retweet partisan content that maximized exposure to partisan talking points. However absurd in principle, campaign coordinators may choose to employ bots talking to bots to boost the perceived popularity of campaign talking points.

The strategic use of bots can, of course, maximize network effects, leading to politically homogeneous communication. While echo chambers are caused by actors in a network influencing each other, it is conceivable that they may have been subjected to microtargeting or other forms of intervention that compound their spatial and associational segregation. Our database of Brexit tweets allowed us to identify whether such ideological clustering was associated with face-to-face, in-person interaction occurring in offline networks (neighbourhood effect) as opposed to being restricted to online networks. Our results showed that echo-chamber communication was largely restricted to neighbouring areas within a 50-km radius, and therefore they likely reproduced political polarization that already existed in offline social networks. The geography of echo chambers was also markedly different for Leave and Remain campaigns, with the former spanning much shorter distances compared with Remain voters.

The analysis of echo-chamber communication in the Leave and Remain subgraphs revealed striking interactions between online activity and geography and substantiated the existence of geographically defined sociopolitical enclaves. The results also identified a geographic patterning in online echo chambers, particularly in the Leave campaign, thus supporting the hypothesis that online echo chambers resulted from conversations that spilled over from in-person interactions. These differences remained significant even when controlling for outliers and abnormalities in the data, with echo chambers in the Leave campaign remaining relatively robust, with peak amplitudes that deviated from the rest of the network and from the distribution observed with the random reshuffling of users' locations.

Troll farms offshore

A key objective of information warfare is to create confusion, disorder, and distrust behind enemy lines through the use of grey or black propaganda. While the Brexit Bots largely impersonated individuals, these operations were reportedly based in Bristol, and were therefore domestic operations typically characterized as white propaganda. There are, however, documented instances of social media manipulation by foreign actors in the context of the Brexit debate, with the IRA, a so-called 'troll factory' reportedly linked to the Russian government, featuring prominently in the set of black propaganda operations rolled out in the course of the referendum campaign.

The IRA is a secretive private company, based in St Petersburg, that engages in information warfare based on fictitious personas created on social media platforms. The IRA is often referred to as a 'troll factory' due to its relentless engagement with social media trolling and the incitement of political discord using fake identities. IRA operations were largely dependent on supervised posting, with employees working 12-hour shifts and managing at least six Facebook fake profiles and ten Twitter fake accounts. The IRA activity was remarkable in its ability to identify and appropriate the subculture or social identity of relatively narrowly defined targets, which in the case of Brexit translated to a sense of economic insecurity experienced by blue-collar working-class voters.

The year 2013 marks the creation of the IRA by Yevgeny Prigozhin in St Petersburg, Russia. It also marks the beginning of IRA's (foreign) black propaganda operations that were pivotal in the Brexit campaign, with 2013 being the year when over one-quarter of the Twitter accounts managed by the IRA became active. Grey propaganda accounts, which also posted Brexit content, appear to be the most complex operation, with one-third of such accounts created in 2013 and a further 42 per cent in 2014. The median activity of grey accounts falls on 29 June 2016, just one week after the UK EU membership referendum. This is similar to the activity patterns of the Brexit Bots, which albeit largely domestically created and managed, were also more active in the period leading up to the referendum and significantly less so in the wake of the vote. The majority of the content posted by the Brexit Bots has disappeared from the internet, comprising 55 per cent of the URLs posted by these accounts.

The content posted by these accounts reveals the intersections between Twitter activity and the broader informational environment around the platform. The content that remained accessible after the vote did not entail outright lies and therefore could not be described as 'fake news'. Instead, the material was rich in rumours, unconfirmed events, and human-interest stories with an emotional and populist appeal. Much like the content seeded by the IRA trolls, the Brexit Bots linked to media sources both mainstream and user-generated, including dubious news stories sourced from self-referencing 'blews'. They also linked to mainstream media that was either quoted out of context, framed to emphasize a partisan aspect of the reportage, or edited with incomplete facts. As such, low-quality information in the Brexit database includes textual or audiovisual content tailored to nudge or interfere with online deliberation by deceiving, confusing, or misinforming.

Misinformation and disinformation campaigns have evolved to maximize the biases of social media platforms, particularly the attention economy and the social media supply chain of engaging rich content. This has led to a shift from narratives exploring how online social networks support gatewatching and citizen journalism to narratives emphasizing how social media reinforce polarization and division in a landscape marked by tribalism and information warfare. The centralized management of the attention economy driving social media platforms is only possible due to the widespread implementation of opaque AI systems to manage interactions and the attention that users devote to content. This critical transformation in the internet social infrastructure was spearheaded by Facebook's algorithmization of communities, leading to the repurposing of user and networked interaction for attention-grabbing campaigns and large-scale disinformation campaigns.

Politics erased

A remarkable feature of the Brexit database is that much of its content has disappeared from the internet. This is consistent with the modus operandi of influence operations that amplify hyperpartisan content using large botnets that disappear after the campaign. Social media platforms are, of course, invested in flagging and removing problematic content and false amplification, but the removal of social media posts and accounts also removes the public accord and public record of the decision-making process underpinning public consultation and deliberation. In this context, research on the politics of deletion implemented by social platforms becomes an exercise in reverse engineering, as content that has been deleted or blocked from social platforms is likely to be problematic content.

Ephemerality is, of course, a vital component of social media communication, but it is not a desirable design for political campaigns. The disappearance of one-third of the discussion underpinning the Brexit referendum is particularly worrying. Indeed, the large-scale removal of social media content has the potential to alter the record of social interactions, preventing forensic analysis and academic research on influence operations. For comparison, the 2019 UK General Election registered a relatively low rate of tweet decay, with only 6.7 per cent of election-related tweets removed and less than 2 per cent of accounts deleted from Twitter. The disappearance of content and user accounts in the course of the Brexit debate is significantly higher though, with the majority of the content posted with some prominent hashtags having disappeared since the vote.

The magnitude of content deletion is astonishing. Nearly 3 million messages posted by 1 million users in the last days of the Brexit referendum campaign have disappeared. Only about half of the most active accounts that tweeted the referendum continue to operate publicly and 20 per cent of all accounts are no longer active. Content deletion is, of course, sensitive to developments on the ground. In the weeks leading up to the Brexit referendum, deletion rose from 19 per cent to 33 per cent. After the referendum, the fraction of deleted tweets peaks from 27 per cent to one-third. Tweet decay decreases in the ensuing months but only becomes consistent with the figures reported in previous studies in the premiership of Boris Johnson. Tweet decay in openly partisan hashtags associated with the Leave campaign is particularly staggering at 42 per cent.

The fraction of deleted tweets and accounts that participated in the Brexit debate sits at 20 per cent, with 155,157 accounts no longer existing and 36,159 blocked by Twitter. Both #voteleave and #voteremain account for a significant share of the hashtags in the data, with #voteleave being significantly more likely to appear in tweets that are no longer available. Indeed, more messages from the Leave campaign have disappeared from Twitter than the entire universe of messages posted on Twitter associated with the Remain campaign. The campaign accounts @ivotestay and @ivoteleave posted 15,928 and 11,647 tweets, respectively, in the 13-day period leading up to the vote, and since the referendum both accounts have been suspended by Twitter. Only about half of the most active accounts in the referendum debate continue to operate publicly, and more worryingly, one-third of all tweets that shaped the discussion about the referendum are no longer retrievable, with around half of these messages having disappeared because the seeding account was removed, blocked, deleted, suspended, or set to private. Suspended accounts were particularly prolific, posting nearly 10 per cent of the entire conversation about the referendum on Twitter.

Our analyses of the Brexit tweets suggest that a sizeable portion of the debate on Twitter was not designed for permanence. It also suggests that deleted tweets may be a proxy for manipulation and disinformation, as much of the deleted content resulted from Twitter actively suspending and ultimately blocking user accounts. This poses considerable challenges for informed public deliberation around matters where the issue being deliberated on is constantly disappearing from public scrutiny. Influence operations are, of course, expertly exploiting these limitations by offloading problematic content that is removed from social media platforms before the relentless news cycle can successfully correct the narratives championed through highly volatile social media content.

The results of the Brexit project suggest that low-quality information shares a repertoire of features that include stylistic and temporal signatures resulting from reduced editorial practices. This content was produced in bulk for quick turnover and characterized by a short shelf life, a crucial point of departure from quality content that requires extensive editorial work at various levels of the news industry and is designed for durability. They were typically hosted by social media platforms and content curation services. The messages were characterized by less-well-edited text, highly partisan discourse, and user-generated content that is quickly modified or erased. Deleted tweets were typically associated with the endorsement of ideological partisanship and exhibited rhetorical features that expressed both certainty and speculation.

Finally, we could not untangle the media effects inadvertently triggered by algorithmic systems from those driven by social media manipulation. Twitter's recommender system was, nonetheless, central to the distribution of polarizing, hyperpartisan Brexit content. These algorithms filter, rank, and select content to be shown to users, often prioritizing content that deepens social divides. Changes implemented to Twitter's Trending Topics are likely to have had tangible effects on the selection and distribution of tabloid-like news content during the Brexit campaign. This social algorithmization of news can reinforce patterns of interaction and the sharing of content in tightly clustered communities that support and reinforce the insularity of users, with the classification of users in communities being processed by imperfect machine learning algorithms that are then built into the system recursively, so that social groups identified by and modelled with digital trace data can take up a life of their own on social platforms, with niche subcommunities of Leave and Remain supporters only partially meeting their physical counterparts.

References

Abokhodair, N., Yoo, D. and McDonald, D. W. 2015. Dissecting a Social Botnet: Growth, Content and Influence in Twitter. *In:* 18th ACM Conference on Computer Supported Cooperative Work and Social Computing, Vancouver, BC, Canada. 2675208: ACM, 839–851.

Acker, A. and Donovan, J. 2019. Data craft: a theory/methods package for critical internet studies. *Information, Communication & Society,* 22, 1590–1609.

Ananny, M. and Crawford, K. 2018. Seeing without knowing: limitations of the transparency ideal and its application to algorithmic accountability. *New Media & Society,* 20, 973–989.

Apuzzo, M. and Santariano, A. 2019. Russia is targeting Europe's elections: so are far-right copycats. *The New York Times,* 12 May. Available from: www.nytimes.com/2019/05/12/world/europe/russian-propaganda-influence-campaign-european-elections-far-right.html [accessed 1 January 2020].

Arif, A., Stewart, L. G. and Starbird, K. 2018. Acting the part: examining information operations within #BlackLivesMatter discourse. *Proceedings of the ACM on Human-Computer Interaction,* 2, 20.

Asthana, A., Quinn, B. and Mason, R. 2016. UK votes to leave EU after dramatic night divides nation. *The Guardian,* 25 June. Available from: www.theguardian.com/politics/2016/jun/24/britain-votes-for-brexit-eu-referendum-david-cameron [accessed 1 January 2020].

Badawy, A., Addawood, A., Lerman, K. and Ferrara, E. 2019. Characterizing the 2016 Russian IRA influence campaign. *Social Network Analysis and Mining,* 9, 31.

Bagdouri, M. and Oard, D. W. 2015. On Predicting Deletions of Microblog Posts. *In: Proceedings of the 24th ACM International Conference on Information and Knowledge Management,* Melbourne, Australia. ACM, 1707–1710.

REFERENCES

Bakshy, E., Messing, S. and Adamic, L. 2015. Exposure to ideologically diverse news and opinion on Facebook. *Science,* 348, 1130–1132.

Barberá, P. 2014. How Social Media Reduces Mass Political Polarization: Evidence from Germany, Spain, and the US. American Political Science Association Conference, San Francisco, CA.

Barberá, P., Jost, J. T., Nagler, J., Tucker, J. A. and Bonneau, R. 2015. Tweeting from left to right: is online political communication more than an echo chamber? *Psychological Science,* 26, 1531–1542.

Bardon, A. 2019. *The Truth about Denial: Bias and Self-deception in Science, Politics, and Religion.* New York: Oxford University Press.

Barthelemy, M. 2014. Spatial Networks. *In*: Alhajj, R. and Rokne, J. (eds) *Encyclopedia of Social Network Analysis and Mining* (pp 1967–1976). New York: Springer.

Bastos, M. 2015. Shares, pins, and tweets: news readership from daily papers to social media. *Journalism Studies,* 16, 305–325.

Bastos, M. 2016. Digital Journalism and Tabloid Journalism. *In*: Franklin, B. and Eldridge, S. (eds) *Routledge Companion to Digital Journalism Studies* (pp 217–225). London: Routledge.

Bastos, M. 2019. Tabloid Journalism. *In: The International Encyclopedia of Journalism Studies* (pp 1–6). John Wiley & Sons. doi: 10.1002/9781118841570.iejs0144

Bastos, M. 2021a. From global village to identity tribes: context collapse and the darkest timeline. *Media and Communication,* 9, 50–58.

Bastos, M. 2021b. Network Spillover Effects and the Dyadic Interactions of Virtual, Social, and Spatial. *In*: Million, A., Haid, C., Ulloa, I. C. and Baur, N. (eds) *Spatial Transformations* (pp. 169–180). London: Routledge.

Bastos, M. 2021c. This account doesn't exist: tweet decay and the politics of deletion in the Brexit debate. *American Behavioral Scientist,* 65, 757–773.

Bastos, M. 2022a. Five challenges in detection and mitigation of disinformation on social media. *Online Information Review,* 46, 413–421.

Bastos, M. 2022b. *Spatializing Social Media: Social Networks Online and Offline*. London: Routledge.

Bastos, M. and Farkas, J. 2019. 'Donald Trump is my president!': The Internet Research Agency propaganda machine. *Social Media + Society*, 5.

Bastos, M. and Mercea, D. 2016. Serial activists: political Twitter beyond influentials and the Twittertariat. *New Media & Society*, 18, 2359–2378.

Bastos, M. and Mercea, D. 2018a. Parametrizing Brexit: mapping Twitter political space to parliamentary constituencies. *Information, Communication & Society*, 21, 921–939.

Bastos, M. and Mercea, D. 2018b. The public accountability of social platforms: lessons from a study on bots and trolls in the Brexit campaign. *Philosophical Transactions of the Royal Society A: Mathematical, Physical and Engineering Sciences*, 376, 20180003.

Bastos, M. and Mercea, D. 2019. The Brexit Botnet and user-generated hyperpartisan news. *Social Science Computer Review*, 37, 38–54.

Bastos, M., Recuero, R. C. and Zago, G. S. 2014. Taking tweets to the streets: a spatial analysis of the Vinegar Protests in Brazil. *First Monday*, 19.

Bastos, M., Mercea, D. and Charpentier, A. 2015. Tents, tweets, and events: the interplay between ongoing protests and social media. *Journal of Communication*, 65, 320–350.

Bastos, M., Mercea, D. and Baronchelli, A. 2018. The geographic embedding of online echo chambers: evidence from the Brexit campaign. *PLOS ONE*, 13, e0206841.

Bastos, M., Mercea, D. and Goveia, F. 2021a. Guy next door and implausibly attractive young women: the visual frames of social media propaganda. *New Media and Society*, 25, 2014–2033.

Bastos, M., Walker, S. and Simeone, M. 2021b. The IMPED Model: detecting low-quality information in social media. *American Behavioral Scientist*, 65, 863–883.

Becker, H. 1949. The nature and consequences of black propaganda. *American Sociological Review*, 14, 221–235.

REFERENCES

Becker, S. O., Fetzer, T. and Novy, D. 2016. *Who voted for Brexit? A Comprehensive District-Level Analysis*. Coventry: Centre for Competitive Advantage in the Global Economy, University of Warwick.

Benkler, Y., Faris, R., Roberts, H. and Zuckerman, E. 2017. Breitbart-led right-wing media ecosystem altered broader media agenda. *Columbia Journalism Review*, 3 March. Available from: www.cjr.org/analysis/breitbart-media-trump-harvard-study.php [accessed 1 January 2020].

Benkler, Y., Faris, R., and Roberts, H. 2018. *Network Propaganda: Manipulation, Disinformation, and Radicalization in American Politics*. Oxford: Oxford University Press.

Bennett, W. L. and Livingston, S. 2018. The disinformation order: disruptive communication and the decline of democratic institutions. *European Journal of Communication*, 33, 122–139.

Bennett, W. L. and Segerberg, A. 2013. *The Logic of Connective Action: Digital Media and the Personalization of Contentious Politics*, Cambridge: Cambridge University Press.

Bertolin, G. 2015. Conceptualizing Russian information operations: info-war and infiltration in the context of hybrid warfare. *IO Sphere*, 10–11.

Bertrand, N. 2017. Twitter will tell Congress that Russia's election meddling was worse than we first thought. *Business Insider*, 30 October.

Bessi, A. and Ferrara, E. 2016. Social bots distort the 2016 US Presidential election online discussion. *First Monday*, 21.

Blank, G. 2017. The digital divide among Twitter users and its implications for social research. *Social Science Computer Review*, 35, 679–697.

Boler, M. and Nemorin, S. 2013. Dissent, Truthiness, and Skepticism in the Global Media Landscape: Twenty-First Century Propaganda in Times of War. *In:* Auerbach, J. and Castronovo, R. (eds) *The Oxford Handbook of Propaganda Studies* (pp 395–417). Oxford: Oxford University Press.

Bolsen, T., Druckman, J. N. and Cook, F. L. 2015. Citizens', scientists', and policy advisors' beliefs about global warming. *The ANNALS of the American Academy of Political and Social Science,* 658, 271–295.

Boxell, L., Gentzkow, M. and Shapiro, J. M. 2017. Greater internet use is not associated with faster growth in political polarization among US demographic groups. *Proceedings of the National Academy of Sciences,* 114, 10612–10617.

Boyd, D. 2008. None of This is Real: Identity and Participation in Friendster. *In*: Karaganis, J. (ed) *Structures of Participation in Digital Culture* (pp 132–157). New York: Social Science Research Council.

Boykoff, M. T. 2008. The cultural politics of climate change discourse in UK tabloids. *Political Geography,* 27, 549–569.

Briant, E. L. 2015. Allies and audiences: evolving strategies in defense and intelligence propaganda. *The International Journal of Press/Politics,* 20, 145–165.

Bright, J. 2016. The social news gap: how news reading and news sharing diverge. *Journal of Communication,* 66, 343–365.

Bruns, A. 2005. *Gatewatching: Collaborative Online News Production.* New York: Peter Lang Publishing Inc.

Bugorkova, O. 2015. Ukraine Conflict: Inside Russia's 'Kremlin Troll Army'. *BBC News.*

Castells, M. 2009. *Communication Power.* Oxford: Oxford University Press.

Castells, M. 2012. *Networks of Outrage and Hope: Social Movements in the Internet Age.* Cambridge: Polity Press.

Celli, F., Stepanov, E. A., Poesio, M. and Riccardi, G. 2016. Predicting Brexit: Classifying Agreement is Better than Sentiment and Pollsters. *In:* Workshop on Computational Modeling of People's Opinions, Personality, and Emotions in Social Media (PEOPLES), Osaka, Japan. The COLING 2016 Organizing Committee, 110–118.

Chadwick, A. 2013. *The Hybrid Media System: Politics and Power.* Oxford: Oxford University Press.

REFERENCES

Cheng, J., Adamic, L. A., Dow, P. A., Kleinberg, J. and Leskovec, J. 2014. Can Cascades Be Predicted? *In:* 23rd International Conference on World Wide Web (WWW'14), Seoul, Korea. ACM, 925–936.

Comor, E. 2001. Harold Innis and 'The bias of communication'. *Information, Communication & Society,* 4, 274–294.

Conover, M. D., Ratkiewicz, J., Francisco, M., Gonçalves, B., Menczer, F. and Flammini, A. 2011. Political Polarization on Twitter. *In:* 5th International AAAI Conference on Weblogs and Social Media (ICWSM11), Barcelona, Spain.

Crouch, C. 1997. The terms of the neo-liberal consensus. *The Political Quarterly,* 68, 352–360.

Cummings, D. 2016. On the Referendum #20: The campaign, physics, and data science – Vote Leave's 'Voter Intention Collection System' (VICS) now available for all. *Dominic Cummings' Blog* [Online]. Available from: https://dominiccummings.wordpress.com/2016/10/29/on-the-referendum-20-the-campaign-physics-and-data-science-vote-leaves-voter-intention-collection-system-vics-now-available-for-all/ [accessed 1 January 2020].

Cunningham, S. B. 2002. *The Idea of Propaganda: A Reconstruction.* Westport, CT: Greenwood Publishing Group.

Dahlgren, P. 2009. *Media and Political Engagement: Citizen, Communication and Press.* Cambridge: Cambridge University Press.

Davis, C. A., Varol, O., Ferrara, E., Flammini, A. and Menczer, F. 2016. Botornot: A System to Evaluate Social Bots. *In: Proceedings of the 25th International Conference Companion on World Wide Web.* International World Wide Web Conferences Steering Committee, 273–274.

De Choudhury, M., Jhaver, S., Sugar, B. and Weber, I. 2016. Social Media Participation in an Activist Movement for Racial Equality. *In:* 10th International AAAI Conference on Web and Social Media, Cologne, Germany: AAAI.

Del Vicario, M., Bessi, A., Zollo, F., Petroni, F., Scala, A., Caldarelli, G. et al 2016. The spreading of misinformation online. *Proceedings of the National Academy of Sciences,* 113, 554–559.

DeVito, M. A. 2017. From editors to algorithms: a values-based approach to understanding story selection in the Facebook news feed. *Digital Journalism,* 5, 753–773.

Doherty, M. 1994. Black propaganda by radio: the German Concordia broadcasts to Britain 1940–1941. *Historical Journal of Film, Radio and Television,* 14, 167–197.

Doherty, M. 2016. Should Making False Statements in a Referendum Campaign Be an Electoral Offence? *Lancashire Law School.* Preston: University of Central Lancashire.

Dorsey, J. P. 2019. We've made the decision to stop all political advertising on Twitter globally. *Twitter* [Online]. Available from: https://twitter.com/jack/status/1189634360472829952 [accessed 30 October 2019].

Dow, P. A., Adamic, L. A. and Friggeri, A. 2013. The Anatomy of Large Facebook Cascades. *In:* 7th International AAAI Conference on Weblogs and Social Media (ICWSM13), 8–11 July, Boston: AAAI.

Dunleavy, P. and Margetts, H. 2001. From majoritarian to pluralist democracy? *Journal of Theoretical Politics,* 13, 295–319.

Dutton, W. H., Reisdorf, B. C., Dubois, E. and Blank, G. 2017. *A Cross-National Survey of Search and Politics* [Online]. Available from: https://ssrn.com/abstract=2944191 [accessed 4 April 2017].

Elections Integrity 2018. Data archive. 17 October: Twitter, Inc.

Ellul, J. 1965. *Propaganda: The Formation of Men's Attitudes.* New York: Knopf.

Expert, P., Evans, T. S., Blondel, V. D. and Lambiotte, R. 2011. Uncovering space-independent communities in spatial networks. *Proceedings of the National Academy of Sciences,* 108, 7663–7668.

Facebook 2018a. Community Standards. Facebook, Inc.

Facebook 2018b. Public Feed API. Facebook, Inc.

Facebook 2018c. Understanding the Facebook: Community Standards Enforcement Report. Facebook, Inc.

Farkas, J. and Bastos, M. 2018. IRA Propaganda on Twitter: Stoking Antagonism and Tweeting Local News. *In:* 9th International Conference on Social Media and Society, Copenhagen, Denmark. 3217929: ACM, 281–285.

REFERENCES

Farkas, J. and Schou, J. 2018. Fake news as a floating signifier: hegemony, antagonism and the politics of falsehood. *Javnost-The Public,* 25, 298–314.

Farkas, J. and Schou, J. 2019. *Post-truth, Fake News and Democracy: Mapping the Politics of Falsehood.* New York: Routledge.

Farkas, J., Schou, J. and Neumayer, C. 2018. Cloaked Facebook pages: exploring fake Islamist propaganda in social media. *New Media & Society,* 20, 1461444817707759.

Ferrara, E. 2017. Disinformation and social bot operations in the run up to the 2017 French presidential election. *First Monday,* 22.

Ferrara, E., Varol, O., Davis, C., Menczer, F. and Flammini, A. 2016. The rise of social bots. *Communications of the ACM,* 59, 96–104.

Festinger, L., Schachter, S. and Back, K. 1950. *Social Pressures in Informal Groups: A Study of Human Factors in Housing.* Michigan: University of Michigan.

Fiegerman, S. and Byers, D. 2017. Facebook, Twitter, Google defend their role in election. *CNN,* 31 October. Available from: http://money.cnn.com/2017/10/31/media/facebook-twitter-google-congress/index.html [accessed 1 January 2020].

Fletcher, R. and Nielsen, R. K. 2017. Using social media appears to diversify your news diet, not narrow it. *NiemanLab,* 21 June. Cambridge, MA: Nieman Foundation for Journalism.

Flew, T. 2019. Guarding the gatekeepers: trust, truth and digital platforms. *Griffith Review,* 64, 94–103.

Freelon, D. 2018. Computational research in the post-API age. *Political Communication,* 35, 665–668.

Freelon, D., McIlwain, C. and Clark, M. 2018. Quantifying the power and consequences of social media protest. *New Media & Society,* 20, 990–1011.

Freelon, D., Bossetta, M., Wells, C., Lukito, J., Xia, Y. and Adams, K. 2020. Black trolls matter: racial and ideological asymmetries in social media disinformation. *Social Science Computer Review,* 40, doi: 10.1177/0894439320914853.

Gamon, M., Basu, S., Belenko, D., Fisher, D., Hurst, M. and König, A. C. 2008. BLEWS: Using Blogs to Provide Context for News Articles. *In:* 2nd International AAAI Conference on Weblogs and Social Media (ICWSM'08), Seattle, WA: AAAI.

Gentzkow, M. and Shapiro, J. M. 2011. Ideological segregation online and offline. *The Quarterly Journal of Economics,* 126, 1799–1839.

Gladwell, M. 2010. Small change: why the revolution will not be tweeted. *The New Yorker,* 4 October. Available from: www.newyorker.com/reporting/2010/10/04/101004fa_fact_gladwell [accessed 1 January 2020].

Gleicher, N. 2019. Removing Coordinated Inauthentic Behavior and Spam from India and Pakistan. Facebook. Available from: https://about.fb.com/news/2019/04/cib-and-spam-from-india-pakistan/ [accessed 15 March 2024].

Gorwa, R. and Guilbeault, D. 2018. Understanding Bots for Policy and Research: Challenges, Methods, and Solutions. *arXiv preprint arXiv:1801.06863.*

Groves, R. M. 2006. Nonresponse rates and nonresponse bias in household surveys. *Public Opinion Quarterly,* 70, 646–675.

Hagey, K. and Horwitz, J. 2021. Facebook tried to make its platform a healthier place. it got angrier instead. *The Wall Street Journal,* 15 September. Available from: www.nytimes.com/2018/02/16/world/europe/prigozhin-russia-indictment-mueller.html [accessed 1 January 2020].

Halpern, D. and Gibbs, J. 2013. Social media as a catalyst for online deliberation? Exploring the affordances of Facebook and YouTube for political expression. *Computers in Human Behavior,* 29, 1159–1168.

Hampton, K. N., Sessions, L. F. and Her, E. J. 2011. Core networks, social isolation and new media. *Information, Communication & Society,* 14, 130–155.

Hanretty, C. 2017. Areal interpolation and the UK's referendum on EU membership. *Journal of Elections, Public Opinion and Parties,* 27, 466–483.

REFERENCES

HERE 2013. Geocoder API Developer's Guide. 6.2.45 ed. Available from: http://documentation.developer.here.com/pdf/geocoding_nlp/6.2.45/Geocoder%20API%20v6.2.45%20Developer's%20Guide.pdf [accessed 4 April 2024].

Hermans, F., Klerkx, L. and Roep, D. 2015. Structural conditions for collaboration and learning in innovation networks: using an innovation system performance lens to analyse agricultural knowledge systems. *The Journal of Agricultural Education and Extension,* 21, 35–54.

Hermida, A. 2010. Twittering the news: the emergence of ambient journalism. *Journalism Practice,* 4, 297–308.

Heymann, P., Koutrika, G. and Garcia-Molina, H. 2007. Fighting spam on social web sites: a survey of approaches and future challenges. *IEEE Internet Computing,* 11, 36–45.

Himelboim, I., McCreery, S. and Smith, M. 2013a. Birds of a feather tweet together: integrating network and content analyses to examine cross-ideology exposure on Twitter. *Journal of Computer-Mediated Communication,* 18, 40–60.

Himelboim, I., Smith, M. and Shneiderman, B. 2013b. Tweeting apart: applying network analysis to detect selective exposure clusters in Twitter. *Communication Methods and Measures,* 7, 195–223.

Hipp, J. R., Corcoran, J., Wickes, R. and Li, T. 2014. Examining the social porosity of environmental features on neighborhood sociability and attachment. *PLOS ONE,* 9, e84544.

Hollander, G. D. 1972. *Soviet Political Indoctrination: Developments in Mass Media and Propaganda since Stalin.* New York: Praeger Publishers.

Horrigan, J., Garrett, K. and Resnick, P. 2004. The internet and democratic debate. *Pew Internet & American Life Project,* Washington, DC.

Howard, P. N. and Hussain, M. 2013. *Democracy's Third Wave.* Oxford: Oxford University Press.

Howard, P. N. and Kollanyi, B. 2016. Bots, #Strongerin, and #Brexit: computational propaganda during the UK-EU Referendum. *SSRN.* Available from: https://papers.ssrn.com/sol3/papers.cfm?abstract_id=2798311 [accessed 15 March 2024].

Huckfeldt, R. R. and Sprague, J. 1995. *Citizens, Politics, and Social Communication: Information and Influence in an Election Campaign.* New York: Cambridge University Press.

Huyen Do, V., Thomas-Agnan, C. and Vanhems, A. 2015. Spatial reallocation of areal data – another look at basic methods. *Revue d'Économie Régionale & Urbaine*, May, 27–58.

Hwang, T., Pearce, I. and Nanis, M. 2012. Socialbots: voices from the fronts. *Interactions,* 19, 38–45.

Inglehart, R. F. and Norris, P. 2016. Trump, Brexit, and the Rise of Populism: Economic Have-nots and Cultural Backlash. *American Political Science Association Annual Meeting.* Philadelphia, USA.

Inglehart, R. F. and Norris, P. 2017. Trump and the xenophobic populist parties. *Perspectives on Politics,* 15, 443–454.

Innis, H. A. 2008. *The Bias of Communication.* Toronto: University of Toronto Press.

Iyengar, S., Sood, G. and Lelkes, Y. 2012. Affect, not ideology: a social identity perspective on polarization. *Public Opinion Quarterly,* 76, 405–431.

Iyengar, S., Lelkes, Y., Levendusky, M., Malhotra, N. and Westwood, S. J. 2019. The origins and consequences of affective polarization in the United States. *Annual Review of Political Science,* 22, 129–146.

Jenkins, H., Ford, S. and Green, J. 2012. *Spreadable Media: Creating Value and Meaning in a Networked Culture.* New York: New York University Press.

Jensen, M. 2018. Russian trolls and fake news: information or identity logics? *Journal of International Affairs,* 71, 115–124.

Jiang, S. and Wilson, C. 2018. Linguistic Signals under Misinformation and Fact-Checking: Evidence from User Comments on Social Media. *In: Proceedings of the ACM on Human-Computer Interaction*, 2(CSCW), article 82.

Johnston, R. and Pattie, C. 2011. Social Networks, Geography and Neighbourhood Effects. *In*: Scott, J. and Carrington, P. J. (eds) *The SAGE Handbook of Social Network Analysis* (pp 301–310). London: SAGE.

Jowett, G. S. and O'Donnell, V. 2014. *Propaganda & Persuasion.* Los Angeles, CA: SAGE.

REFERENCES

Kahan, D. M., Jenkins-Smith, H. and Braman, D. 2011. Cultural cognition of scientific consensus. *Journal of Risk Research,* 14, 147–174.

Karlova, N. A. and Fisher, K. E. 2013. A social diffusion model of misinformation and disinformation for understanding human information behaviour. *Information Research,* 18.

Khamis, S., Gold, P. B. and Vaughn, K. 2013. Propaganda in Egypt and Syria's Cyberwars: Contexts, Actors, Tools, and Tactics. *In*: Auerbach, J. and Castronovo, R. (eds) *The Oxford Handbook of Propaganda Studies* (pp 418–438). New York: Oxford University Press.

Kim, Y. M. 2011. The contribution of social network sites to exposure to political difference: the relationships among SNSs, online political messaging, and exposure to cross-cutting perspectives. *Computers in Human Behavior,* 27, 971–977.

Kim, Y. M., Hsu, J., Neiman, D., Kou, C., Bankston, L., Kim, S. Y. et al 2018. The stealth media? Groups and targets behind divisive issue campaigns on Facebook. *Political Communication,* 35, 515–541.

King, G., Pan, J. and Roberts, M. E. 2017. How the Chinese government fabricates social media posts for strategic distraction, not engaged argument. *American Political Science Review,* 111, 484–501.

Krasodomski-Jones, A. 2016. *Talking to Ourselves? Political Debate Online and the Echo Chamber Effect.* London: DEMOS.

Kriesi, H. and Frey, T. 2008. The United Kingdom: Moving Parties in a Stable Configuration. *In*: Grande, E., Kriesi, H., Dolezal, M., Lachat, R., Bornschier, S. and Frey, T. (eds) *West European Politics in the Age of Globalization* (pp 183–207). Cambridge: Cambridge University Press.

Kriesi, H., Grande, E., Lachat, R., Dolezal, M., Bornschier, S. and Frey, T. 2008. Globalization and its Impact on National Spaces of Competition. *In*: Grande, E., Kriesi, H., Dolezal, M., Lachat, R., Bornschier, S. and Frey, T. (eds) *West European Politics in the Age of Globalization* (pp 3–22). Cambridge: Cambridge University Press.

Kruspe, A., Häberle, M., Hoffmann, E. J., Rode-Hasinger, S., Abdulahhad, K. and Zhu, X. X. 2021. Changes in Twitter geolocations: insights and suggestions for future usage. *arXiv preprint arXiv:2108.12251.*

Kuklinski, J. H., Quirk, P. J., Jerit, J., Schwieder, D. and Rich, R. F. 2000. Misinformation and the currency of democratic citizenship. *Journal of Politics,* 62, 790–816.

Kumar, S., Cheng, J., Leskovec, J. and Subrahmanian, V. S. 2017. An Army of Me: Sockpuppets in Online Discussion Communities. *In: Proceedings of the 26th International Conference on World Wide Web.* Perth, Australia: International World Wide Web Conferences Steering Committee, 857–866.

Laclau, E. 2005. *On Populist Reason.* London: Verso.

Langley, P. and Leyshon, A. 2017. Platform capitalism: the intermediation and capitalisation of digital economic circulation. *Finance and Society,* 3, 11–31.

Laniado, D., Volkovich, Y., Scellato, S., Mascolo, C. and Kaltenbrunner, A. 2017. The impact of geographic distance on online social interactions. *Information Systems Frontiers,* 1–16.

Latour, B. 2004. Why has critique run out of steam? From matters of fact to matters of concern. *Critical Inquiry,* 30, 225–248.

Lazarsfeld, P. and Merton, R. 1954. Friendship as a social process: a substantive and methodological analysis. *Freedom and Control in Modern Society,* 18, 18–66.

Lewandowsky, S., Ecker, U. K. and Cook, J. 2017. Beyond misinformation: understanding and coping with the 'post-truth' era. *Journal of Applied Research in Memory and Cognition,* 6, 353–369.

Lewis, J., Cushion, S. and Thomas, J. 2005. Immediacy, convenience or engagement? An analysis of 24-hour news channels in the UK. *Journalism Studies,* 6, 461–477.

Liben-Nowell, D., Novak, J., Kumar, R., Raghavan, P. and Tomkins, A. 2005. Geographic routing in social networks. *Proceedings of the National Academy of Sciences,* 102, 11623–11628.

Lijffijt, J., Nevalainen, T., Säily, T., Papapetrou, P., Puolamäki, K. and Mannila, H. 2016. Significance testing of word frequencies in corpora. *Literary and Linguistic Computing,* 31, 374–397.

REFERENCES

Linebarger, P. 1948. *Psychological Warfare*. Washington, DC: Infantry Journal Press.

Linvill, D. L. and Warren, P. L. 2020. Troll factories: manufacturing specialized disinformation on Twitter. *Political Communication*, 37, 447–467.

MacFarquhar, N. 2018. Yevgeny Prigozhin, Russian oligarch indicted by U.S., is known as 'Putin's Cook'. *The New York Times*, 16 February. Available from: www.nytimes.com/2018/02/16/world/europe/prigozhin-russia-indictment-mueller.html [accessed 1 January 2020].

Marwick, A. E. and Boyd, D. 2011. I tweet honestly, I tweet passionately: Twitter users, context collapse, and the imagined audience. *New Media & Society,* 13, 114–133.

Massanari, A. 2015. #Gamergate and The Fappening: how Reddit's algorithm, governance, and culture support toxic technocultures. *New Media & Society*, 19, 329–346.

Maynard, A. D. 2015. Why we need risk innovation. *Nature Nanotechnology,* 10, 730–731.

McAndrew, F. T. 2017. *The SAGE Encyclopedia of War: Social Science Perspectives*. Thousand Oaks, CA: SAGE.

McCreadie, R., Soboroff, I., Lin, J., Macdonald, C., Ounis, I. and McCullough, D. 2012. On Building a Reusable Twitter Corpus. *In: Proceedings of the 35th International ACM SIGIR Conference on Research and Development in Information Retrieval*. ACM, 1113–1114.

McGarty, C., Thomas, E. F., Lala, G., Smith, L. G. and Bliuc, A. M. 2014. New technologies, new identities, and the growth of mass opposition in the Arab Spring. *Political Psychology,* 35, 725–740.

McLaren, L. M. 2002. Public support for the European Union: cost/benefit analysis or perceived cultural threat? *The Journal of Politics,* 64, 551–566.

McNally, N., and M. T. Bastos. 2023. Auditing Facebook Algorithms: The Elapsed Effects of Facebook News Feed to Engagement with *Guardian* Articles. *In: AoIR Selected Papers of Internet Research*.

McPherson, M. and Smith-Lovin, L. 1987. Homophily in voluntary organizations: status distance and the composition of face-to-face groups. *American Sociological Review,* 52, 370–379.

McPherson, M., Smith-Lovin, L. and Cook, J. M. 2001. Birds of a feather: homophily in social networks. *Annual Review of Sociology,* 27, 415–444.

Mercea, D. and Bastos, M. 2016. Being a serial transnational activist. *Journal of Computer-Mediated Communication,* 21, 140–155.

Messing, S. and Westwood, S. J. 2014. Selective exposure in the age of social media. *Communication Research,* 41, 1042–1063.

Metaxas, P. and Mustafaraj, E. 2010. From Obscurity to Prominence in Minutes: Political Speech and Real-Time Search. *In: Proceedings of the WebSci10: Extending the Frontiers of Society On-Line*, Raleigh, NC.

Metcalf, J. and Crawford, K. 2016. Where are human subjects in big data research? The emerging ethics divide. *Big Data & Society,* 3, 2053951716650211.

Miller, N. 2017. British politicians seek answers in suspected Russian role in Brexit campaign. *The Sydney Morning Herald.* Available from: https://web.archive.org/web/20180319104536/https://www.smh.com.au/world/british-politicians-seek-answers-in-suspected-russian-role-in-brexit-campaign-20171025-gz7lyk.html [accessed 1 January 2020].

Miller, W. L. 1977. *Electoral Dynamics in Britain since 1918*. London: Macmillan.

Mok, D., Wellman, B. and Carrasco, J. 2010. Does distance matter in the age of the Internet? *Urban Studies,* 47, 2747–2783.

Moore, M. 2016. *Tech Giants and Civic Power*. London: Centre for the Study of Media, Communication & Power.

Morozov, E. 2013. *To Save Everything, Click Here: The Folly of Technological Solutionism*. Public Affairs.

Mudde, C. 2000. *The Ideology of the Extreme Right*. Manchester: Manchester University Press.

Mudde, C. 2004. The populist zeitgeist. *Government and Opposition,* 39, 541–563.

REFERENCES

Mudde, C. and Rovira Kaltwasser, C. 2012. *Populism in Europe and the Americas*. Cambridge: Cambridge University Press.

Murthy, D. 2013. *Twitter: Social Communication in the Twitter Age*. Cambridge: Polity.

Murthy, D., Powell, A. B., Tinati, R., Anstead, N., Carr, L., Halford, S. J. et al 2016. Automation, algorithms, and politics| bots and political influence: a sociotechnical investigation of social network capital. *International Journal of Communication,* 10, 20.

Napoli, P. and Caplan, R. 2017. Why media companies insist they're not media companies, why they're wrong, and why it matters. *First Monday*, 22.

Nilsson, B. and Carlsson, E. 2014. Swedish politicians and new media: democracy, identity and populism in a digital discourse. *New Media & Society,* 16, 655–671.

Onnela, J.-P., Arbesman, S., González, M. C., Barabási, A.-L. and Christakis, N. A. 2011. Geographic constraints on social network groups. *PLOS ONE,* 6, e16939.

ONS Geography 2011. National statistics postcode lookup UK. Office for National Statistics. London: ONS.

ONS Geography 2017. NHS postcode directory user guide. *National Statistics Postcode Lookup UK*. Office for National Statistics. Available from: https://data.gov.uk/dataset/national-statistics-postcode-lookup-uk [accessed 4 April 2024].

Owen, L. H. 2019. It is still incredibly easy to share (and see) known fake news about politics on Facebook. *NiemanLab*, 8 November. Cambridge, MA: Nieman Foundation for Journalism.

Papacharissi, Z. 2008. The Virtual Sphere 2.0: The Internet, the Public Sphere, and Beyond. *In*: Chadwick, A. (ed) *Routledge Handbook of Internet Politics* (pp 246–261). New York: Routledge.

Pariser, E. 2012. *The Filter Bubble: What the Internet Is Hiding from You*. London: Penguin.

Parker, G. 2016. Nigel Farage: Eurosceptic scourge of the 'political elite'. *Financial Times*, 12 December. Available from: www.ft.com/content/ab8b3b98-be3a-11e6-8b45-b8b81dd5d080 [accessed 1 January 2020].

Paul, C. and Matthews, M. 2016. The Russian 'firehose of falsehood' propaganda model: why it might work and options to counter it. *Rand Corporation*, 2–7.

Pérez-Rosas, V., Kleinberg, B., Lefevre, A. and Mihalcea, R. 2017. Automatic detection of fake news. *In: Proceedings of the 27th International Conference on Computational Linguistics*, Santa Fe, NM, 3391–3401.

Pew Research Center. *The Demographics of Social Media Users, 2012.* Pew Research Center's Internet & American Life Project, 2013.

Piketty, T. 2014. *Capital in the Twenty-first Century*. Cambridge, MA: The Belknap Press of Harvard University Press.

Preciado, P., Snijders, T. A., Burk, W. J., Stattin, H. and Kerr, M. 2012. Does proximity matter? Distance dependence of adolescent friendships. *Social Networks,* 34, 18–31.

R Development Core Team 2014. R: A Language and Environment for Statistical Computing. *R Foundation for Statistical Computing.* 3.0.3 ed. Vienna, Austria: CRAN.

Ratkiewicz, J., Conover, M., Meiss, M., Gonçalves, B., Patil, S., Flammini, A. et al 2011a. Truthy: Mapping the Spread of Astroturf in Microblog Streams. *In:* 20th International Conference Companion on World Wide Web, Hyderabad, India. ACM, 249–252.

Ratkiewicz, J., Conover, M. D., Meiss, M., Gonçalves, B., Flammini, A. and Menczer, F. 2011b. Detecting and Tracking Political Abuse in Social Media. *In:* 5th International AAAI Conference on Weblogs and Social Media (ICWSM11), Barcelona, Spain.

Rauchfleisch, A. and Kaiser, J. 2020. The false positive problem of automatic bot detection in social science research. *PLOS ONE,* 15, e0241045.

Raymond, J. 2003. *Pamphlets and Pamphleteering in Early Modern Britain*. Cambridge: Cambridge University Press.

Rennie Short, J. 2016. The geography of Brexit: what the vote reveals about the disunited Kingdom. *The Conversation,* 27 June. Available from: http://theconversation.com/the-geography-of-brexit-what-the-vote-reveals-about-the-disunited-kingdom-61633 [accessed 30 April 2024].

REFERENCES

Resnick, P. and Varian, H. R. 1997. Recommender systems. *Communications of the ACM,* 40, 56–58.

Riffe, D., Aust, C. F. and Lacy, S. R. 1993. The effectiveness of random, consecutive day and constructed week sampling in newspaper content analysis. *Journalism & Mass Communication Quarterly,* 70, 133–139.

Rose, K. and McGrory, C. 2016. UK social media statistics for 2016. Available from: www.rosemcgrory.co.uk/2016/01/04/social-media-statistics-2016/ [accessed 29 April 2024].

Rosenberg, M. 2018. Professor apologizes for helping Cambridge Analytica harvest Facebook data. *The New York Times,* 22 April. Available from: www.nytimes.com/2018/04/22/business/media/cambridge-analytica-aleksandr-kogan.html [accessed 1 January 2020].

Rosner, B. 1983. Percentage points for a Generalized ESD many-outlier procedure. *Technometrics,* 25, 165–172.

Rowe, D. 2011. Obituary for the newspaper? Tracking the tabloid. *Journalism,* 12, 449–466.

Rucker, D. D., Tormala, Z. L., Petty, R. E. and Briñol, P. 2014. Consumer conviction and commitment: an appraisal-based framework for attitude certainty. *Journal of Consumer Psychology,* 24, 119–136.

Sanders, M. L. and Taylor, P. M. 1982. *British Propaganda during the First World War, 1914–18.* London, UK: Macmillan International Higher Education.

Schudson, M. 1995. *The Power of News.* Cambridge, MA: Harvard University Press.

Seddon, M. 2014. Documents Show How Russia's Troll Army Hit America. BuzzFeed News [Online]. Available from: www.buzzfeed.com/maxseddon/documents-show-how-russias-troll-army-hit-america [accessed 3 June 2018].

Selivanov, D. 2016. text2vec: Modern Text Mining Framework for R. 0.4.0 ed. Vienna, Austria: CRAN.

Shane, S. and Mazzetti, M. 2018. Inside a 3-Year Russian campaign to influence US voters. *The New York Times,* 16 February.

Shirky, C. 2008. *Here Comes Everybody: The Power of Organizing Without Organizations*. New York: Penguin.

Shorey, S. and Howard, P. N. 2016. Automation, algorithms, and politics | automation, big data and politics: a research review. *International Journal of Communication,* 10, 5032–5055.

Silva, S. 2016. Trump's Twitter Debate Lead Was 'Swelled by Bots'. BBC News [Online]. Available from: www.bbc.co.uk/news/technology-37684418 [accessed 15 December 2016].

Sloan, L. 2017. Who tweets in the United Kingdom? Profiling the Twitter population using the British Social Attitudes Survey 2015. *Social Media + Society,* 3, 2056305117698981.

Sloan, L., Morgan, J., Housley, W., Williams, M., Edwards, A., Burnap, P. et al 2013. Knowing the tweeters: deriving sociologically relevant demographics from Twitter. *Sociological Research Online,* 18, 7.

Starbird, K., Maddock, J., Orand, M., Achterman, P. and Mason, R. M. 2014. Rumors, False Flags, and Digital Vigilantes: Misinformation on Twitter after the 2013 Boston Marathon Bombing. *In: IConference 2014 Proceedings,* 654–662.

Starbird, K., Arif, A. and Wilson, T. 2019. Disinformation as Collaborative Work: Surfacing the Participatory Nature of Strategic Information Operations. *In: Proceedings of the ACM on Human-Computer Interaction,* 3, 1–26.

Statista. 2021. Percentage of US Adults Who Use Twitter as of February 2021, by Age Group. *Statista,* https://www.statista.com/statistics/265647/share-of-us-internet-users-who-use-twitter-by-age-group/

Stewart, L. G., Arif, A., Nied, A. C., Spiro, E. S. and Starbird, K. 2017. Drawing the Lines of Contention: Networked Frame Contests Within #BlackLivesMatter Discourse. In: *Proceedings of the ACM on Human-Computer Interaction,* 1, 96.

Storper, M. 2018. Separate worlds? Explaining the current wave of regional economic polarization. *Journal of Economic Geography,* 18, 247–270.

REFERENCES

Subrahmanian, V. S., Azaria, A., Durst, S., Kagan, V., Galstyan, A., Lerman, K. et al 2016. The DARPA Twitter Bot challenge. *Computer,* 49, 38–46.

Sunstein, C. R. 2007. *Republic.com 2.0.* Princeton, NJ: Princeton University Press.

Swinford, S. 2016. Britain could be up to £70 billion worse off if it leaves the Single Market after Brexit, IFS warns. *The Telegraph.* London.

Tajfel, H. 1974. Social identity and intergroup behavior. *Social Science Information,* 13, 65–93.

Tajfel, H. 1978. Social Categorization, Social Identity and Social Comparison. *In*: Tajfel, H. (ed) *Differentiation Between Social Groups: Studies in the Social Psychology of Intergroup Relations* (pp 61–76). London: Academic Press.

Takhteyev, Y., Gruzd, A. and Wellman, B. 2012. Geography of Twitter networks. *Social Networks,* 34, 73–81.

Taylor, P. M. 2003. *Munitions of the Mind: A History of Propaganda from the Ancient World to the Present Era.* Manchester: Manchester University Press.

Temple Lang, D. 2016. RCurl: General Network (HTTP/FTP/...) Client Interface for R. 1.95–4.8 ed. Vienna, Austria: CRAN.

Terren, L. and Borge, R. 2021. Echo chambers on social media: a systematic review of the literature. *A Review of Communication Research.*

The Economist 2018. Russian disinformation distorts American and European democracy. *The Economist.*

Thurman, N., Lewis, S. C. and Kunert, J. 2019. Algorithms, automation, and news. *Digital Journalism,* 7, 980–992.

Tufekci, Z. and Wilson, C. 2012. Social media and the decision to participate in political protest: observations from Tahrir Square. *Journal of Communication,* 62, 363–379.

Turner, B. 2002. *Orientalism, Postmodernism and Globalism.* Singapore: Routledge.

Twitter 2017a. Automation Rules. Twitter, Inc.

Twitter 2017b. November 24. *Letter from Nick Pickles, Twitter, to the Digital, Culture, Media and Sport Committee* [Letter to Culture Digital.

Twitter. 2018a, January 19. *Letter from Nick Pickles, Twitter, to the Chair of Fake News Inquiry* [Letter to Chair of Fake News Inquiry.

Twitter 2018b. Retweet FAQs. Twitter, Inc.

Twitter 2018c. Twitter Privacy Policy. Twitter, Inc.

Twitter 2018d. Update on Twitter's review of the 2016 US Election. (Global Public Policy).

Twitter 2019a. More about restricted uses of the Twitter APIs. https://developer.twitter.com/en/developer-terms/more-on-restricted-use-cases.html

Twitter. 2019b. *Private communication: Only information available in the API can be made public* [Letter to M. T. Bastos].

Tyler, M., Grimmer, J., and Iyengar, S. (2022). Partisan Enclaves and Information Bazaars: Mapping Selective Exposure to Online News. *The Journal of Politics*, 84, 1057–1073.

United States Senate Committee 2017. Testimony of Sean J. Edgett, Acting General Counsel, Twitter, Inc., to the United States Senate Committee on the Judiciary, Subcommittee on Crime and Terrorism. https://www.judiciary.senate.gov/imo/media/doc/10-31-17%20Edgett%20Testimony.pdf

US District Court 2018. *United States of America versus Internet Research Agency LLC*. Washington, DC: United States District Court for the District of Columbia.

Vaccari, C., Valeriani, A., Barberá, P., Jost, J. T., Nagler, J. and Tucker, J. A. 2016. Of echo chambers and contrarian clubs: exposure to political disagreement among German and Italian users of Twitter. *Social Media + Society*, 2, 2056305116664221.

Vallis, O., Hochenbaum, J., Kejariwal, A., Rudis, B. and Tang, Y. 2014. AnomalyDetection: Anomaly Detection Using Seasonal Hybrid Extreme Studentized Deviate Test. *R Package Version*. Available from: https://www.rdocumentation.org/packages/AnomalyDetection/versions/1.0 [accessed 30 April 2024].

Varol, O., Ferrara, E., Davis, C. A., Menczer, F. and Flammini, A. 2017. Online Human-Bot Interactions: Detection, Estimation, and Characterization. *In:* 11th International AAAI Conference on Weblogs and Social Media, Montréal, Canada. eprint arXiv:1703.03107: AAAI, 280–289.

REFERENCES

Vinhas, O. and Bastos, M. 2022. Fact-checking misinformation: eight notes on consensus reality. *Journalism Studies,* 23, 448–468.

Vosoughi, S., Roy, D. and Aral, S. 2018. The spread of true and false news online. *Science,* 359, 1146–1151.

Walker, S. and #FAIL! workshop. (2015). *The Complexity of Collecting Social Media Data in Ephemeral Contexts.* Internet Research 16, Phoenix, AZ.

Walker, S., Mercea, D. and Bastos, M. 2019. The disinformation landscape and the lockdown of social platforms. *Information, Communication and Society,* 22, 1531–1543.

Weedon, J., Nuland, W. and Stamos, A. 2017. Information Operations and Facebook. Facebook.

Welch, D. 2013. *Propaganda, Power and Persuasion: From World War I to Wikileaks.* London: I.B.Tauris.

Wills, J. 2015. Populism, localism and the geography of democracy. *Geoforum,* 62, 188–189.

Wojcieszak, M. 2010. 'Don't talk to me': effects of ideologically homogeneous online groups and politically dissimilar offline ties on extremism. *New Media & Society,* 12, 637–655.

Wojcieszak, M. and Mutz, D. C. 2009. Online groups and political discourse: do online discussion spaces facilitate exposure to political disagreement? *Journal of Communication,* 59, 40–56.

Wong, L. H., Pattison, P. and Robins, G. 2006. A spatial model for social networks. *Physica A: Statistical Mechanics and its Applications,* 360, 99–120.

Woods, D. 2009. Pockets of resistance to globalization: the case of the Lega Nord. *Patterns of Prejudice,* 43, 161–177.

Woolley, S. C. 2016. Automating power: social bot interference in global politics. *First Monday,* 21.

Woolley, S. C. and Howard, P. N. 2016. Automation, algorithms, and politics | political communication, computational propaganda, and autonomous agents – introduction. *International Journal of Communication,* 10.

Xu, J.-M., Burchfiel, B., Zhu, X. and Bellmore, A. 2013. An Examination of Regret in Bullying Tweets. *In: Proceedings of the 2013 Conference of the North American Chapter of the Association for Computational Linguistics: Human Language Technologies*, 697–702.

Youmans, W. L. and York, J. C. 2012. Social media and the activist toolkit: user agreements, corporate interests, and the information infrastructure of modern social movements. *Journal of Communication*, 62, 315–329.

Zheng, X., Lai, Y. M., Chow, K. P., Hui, L. C. K. and Yiu, S. M. 2011. Sockpuppet Detection in Online Discussion Forums. *In:* 7th International Conference on Intelligent Information Hiding and Multimedia Signal Processing, 14–16 October, Dalian, China, 374–377.

Zimmer, M. 2010. 'But the data is already public': on the ethics of research in Facebook. *Ethics and Information Technology*, 12, 313–325.

Zimmer, M. 2015. The Twitter archive at the Library of Congress: challenges for information practice and information policy. *First Monday*, 20.

Zollo, F., Bessi, A., Del Vicario, M., Scala, A., Caldarelli, G., Shekhtman, L. et al 2017. Debunking in a world of tribes. *PLOS ONE*, 12, e0181821.

Zuckerberg, M. 2017. Building Global Community. Facebook.

Zuckerman, E. 2014. New media, new civics? *Policy & Internet*, 6, 151–168.

Index

#15maydebout 128
#betteroffout 129, 132
#blacklivesmatter 128
#brexit 127
#voteleave 125, 129, 132, 157
#voteremain 125, 157
@- mentions
  bots 70, 73
  distance travelled calculations 61
  echo chambers 60, 61
  identification of affiliation 18
  troll factories 96
@brndstr 101
@EUFear 80
@evangeliney0ung 102
@ivoteleave 123, 157
@ivotestay 123, 157
@nero 75, 78
@no_eusssr_thx 80
@stevemmensUKIP 75–76, 80
@trendingpls 80, 83
@uk5am 80
@vote_leave 83, 87

A

Abokhodair, Norah 77
accountability 136–147
advertising 107–108, 122, 144
affective polarization 50, 144
affective validation 144
affiliation 7, 13, 17–20, 23, 40, 50, 52, 59, 98
affordances 105
affordances of social platforms 52, 105, 109, 119, 121, 137, 141, 152
age 29, 149
AI (Artificial Intelligence) 100, 106, 156
algorithms
  algorithmic auditing 112
  algorithmic filtering 30, 53
  algorithmic junk 144–147, 158–159
  algorithmization of communities 107–108, 146, 156
  deep learning algorithms 36, 39
  filter bubbles 49, 52–55, 152
  hidden nature of 111, 144–147
  machine learning algorithms 20, 34, 36–41, 42, 71, 146
  and networked publics 106–108
alienation from electoral processes 110, 111
altering the record 119, 132, 136, 156
alt-right activism 75
anomaly detection 15–16
anonymity 93
Application Programming Interfaces (APIs) 12, 13, 15, 16, 17, 76, 108–110, 121, 123, 125
  see also REST API; Streaming API
archives of disinformation 108, 119
Article 50 trigger 131, 135
attention economy 104, 156
authoritarianism 93
automated accounts 96, 118, 120, 150
  see also bots
automated activity 77, 149
automatic posting 69–70, 150
average number of tweets per user 8, 9

B

Bagdouri, Mossaab 122, 123
Ball, James 99

banned accounts 97
Barberá, Pablo 30, 53, 54, 55–56
Bastos, Marco 12, 20, 36, 44, 50, 52, 58, 74, 75, 80, 93, 96, 97, 99, 101, 105, 106, 109, 118, 120, 128, 129, 131, 137, 140, 141, 142, 143, 144, 145, 146, 147
Becker, Howard 92, 94
Becker, Sascha O. 27, 29, 36, 42
Benkler, Yochai 52, 82, 145–146
Bennett, W. Lance 105, 110, 111
Bessi, Alessandro 71, 72, 73, 74, 93, 110
bias 51, 104, 144
black propaganda 92, 93–94, 96, 98–99, 154–155
bleus 82, 102, 155
blocked content 109
blocking of accounts 10, 77, 101, 121, 124, 127, 143, 156, 158
blue-collar workers 28, 29, 34, 35, 36, 149, 154
BNP (British National Party) 42
Boler, Megan 92
'bombs' 74
borders 35
'Bot Studio for Brands' 101
Botometer 76
bots
  bot activity 69–87, 150–154
  bot detection 12, 20, 70–72, 75–76, 150–151
  bot to bot communication 82–87
  botmasters 16, 71, 75, 77, 153
  botnets 16, 17, 70, 78, 139, 156
  post-referendum activity 87
  prevalence of 74
  reach versus targeted messages 81
  removal 76–77
  retweet cascades 81, 82–87
  shelf life of content 81–82
  specialization 81
  and superusers 69–76
  timing of activity 80, 87
  tweet decay 122
  Twitter 'bombs' 74
Brexit Botnet 69, 75–76, 80, 151

Brexit Bots 16, 17, 79–82, 83, 93–103, 140, 152, 154, 155
Brexit Classifier algorithm 38–41
Brexit value space 31–33
Bristol troll farms 99–103, 154
British Twitter Monthly Active Userbase (BTMAU) 1, 7–12, 131, 148–149
broken links 101
business models (social media) 105, 106–107
Buzzfeed 99, 145

C

Cambridge Analytica 6, 104, 107, 108
campaign advocacy 18–20
campaign period 12
capitalization of digital circulation 53
cascades *see* retweet cascades
Castells, Manuel 93, 105, 106
Celli, Fabio 31, 38, 79
censorship 93
Chadwick, Andrew 105
Chequers Plan 131–132
citizen journalism 105–106, 155
civic participation 136, 147
Classifier algorithm 38–41
clickbait 146
clicktivism 49
climate change 51
cloud-based application interfaces 106, 146
collectivity 105, 136–137
commercialization of previously public spaces 106–108
commodification of digital data 53
community building 107, 146
Community Standards 118, 120, 140
confirmation bias 51
consensus-building processes 33, 51
conservative ideologies 51, 56
Conservative Party
  nationalist rhetoric 28, 30, 149

INDEX

traditional conservative values 30–31, 35
working classes 31
conspiracy theories 98
constituency size 45
constituency-level data 41, 44
constructed week sampling 15
contagion effects 57–58
content curation services 82, 142–143, 151–152, 158
cosmopolitan values 19, 36
council wards 7, 21–22
cross-bubble tweets 20, 59, 61, 62, 64–66
cross-partisan communication 59
CrowdTangle 108
cultural backlash 28–29, 35, 36, 80
cultural conservativism 31
cultural realignment 28, 29
Cummings, Dominic 31, 79

D

daisy chaining 112, 120
data access issues in research 108–110
data analytics 31, 79
data collection pipeline 12–18
data lockdowns 104, 108
Davis, Clayton 76
dead links 101–102
decay *see* tweet decay
decentralization of media 92–93, 106
deep learning algorithms 36, 39
deleted accounts 76–77, 94, 99, 101, 102, 117–135, 137, 138, 150–151, 156–159
deleted content 100–102, 117–135, 140–141
deleted link targets 101–102
deleted tweets 16, 70, 100–101, 102, 109, 117–135, 137, 142, 155, 156–159
deletion rates 102, 122–123, 129–131, 139
democratization of public discourse 105, 144
demographic data 7, 10, 149
density of tweets 9
detection
  bot detection 12, 20, 70–72, 75–76, 150–151
  problematic content 141–144
digital forensics 108
digital networked politics 105
digital trace data 146, 152, 159
disappearing content 100, 109, 156–157
  see also deleted content
disinformation 74–75, 93, 104–113, 119, 121, 139, 141, 142, 155–156, 158
distance travelled
  calculations 60–61
distributed networks services 107
district-level data 7–8, 21–22, 41
Document-Term Matrix (DTM) 40
dormant accounts 99
Dubai 101
due process of law 33

E

echo chambers 14, 20, 30, 49, 52–68, 152, 153
economics
  Brexit Classifier algorithm 39–41
  in Brexit tweets 42–46
  economic insecurity hypothesis 28–29, 35
  economism 20, 33, 37, 38, 39–41, 42–46
  hard to split economism from populism 41
  Leave campaign 150
  regional distribution 45
EconPop variable 42–46
Elections Integrity 97, 108, 119
electoral districts 8, 21–22, 41
elites 32, 35–36, 37, 95, 96, 106, 149
Ellul, Jacques 92
Emmens, Steve 75–76
end-to-end encryption 107

Enterprise 124
environmental protection 29, 35
ephemerality 12, 112, 119, 129–135, 137–139, 141, 142, 156–157, 158
equality 29
ethical considerations 16
ethnic values 149–150
European integration 35, 36
European Parliament elections (2014) 42
evidence-based policymaking 33
experts 33, 51–52

F

Facebook
  algorithms 106, 108, 144–147, 156
  Cambridge Analytica 6, 104, 107, 108
  Community Standards 120, 140
  filter bubbles 54
  gatekeeping 145
  influence operations 119–120
  IRA troll farms 75, 151
  Metaverse 107
  News Feed 106, 144, 145
  profiling technology 100
  Public Feed API 110
  user privacy 5
fact-checking agencies 111
fake accounts 70, 95, 96, 118, 140
fake identities 95, 98, 150
fake news 82, 101, 111, 118, 139, 144, 155
false amplification 70, 118–119, 120, 137, 139
false consensus 72
false negatives 76, 124
'false news' 139
false positives 17, 19, 75–76, 124
Farkas, Johan 93, 96, 97, 101, 106, 128, 139
far-right material 82
fearmongering 98
feedback loops 57

Ferrara, Emilio 72, 73, 74, 77, 93, 110
filter bubbles 49, 52–55, 152
filtering approach to bot detection 76, 78
'firehose of falsehood' model 111, 113, 120
flagged content 111, 112, 113, 118, 121
forensic analysis 108, 119, 120, 156
France, 2017 elections 110
Frey, Timotheos 30, 31, 35

G

gaslighting 108–110, 137
gatekeeping 106, 107, 145
gatewatching 105–106, 155
gender 10, 80
generalizability 15
geographic information
  and echo chambers 58
  geocoding 13–15, 21–23
  geographic distribution of BTMAU 10, 11, 38, 40, 44, 149, 153
  geographic location of users 21–23
  geographic propinquity 50, 55–68
  geographical social networks 50
  geolocation data 7
  granularity of geographic data 21
  sender-receiver geographical distance 14
  social ties 56
  and spatial segregation 54
  triangulation 7, 13–15
geo-spatial social networks 61, 63, 67
Gladwell, Malcolm 49
global communities 107
global media ecology 105
globalism 20, 31, 32, 37–41, 42–46, 127
globalization 29, 32, 36
GlobNat variable 42–46
Google 145, 147

governance structures 5–6, 52, 106, 107
grammar/spelling 143
graph-based approaches to bot detection 76–79
grey propaganda 92, 93–94, 96, 98, 99, 155
group identities 55–56, 96–97, 152
GSS (Government Statistical Service) 22

H

Hagey, Keach 145
hashtags
  bots 71
  data collection 12, 15
  and deep learning algorithms 39
  deleted content 125, 127, 128–129, 130, 131, 132, 157
  hashtag stuffing 143
  Leave campaign 127, 128, 131, 132
  multiple 18
  openly partisan hashtags 128, 132, 157
  post-referendum activity 132
  proxy for ideological affiliation 18, 42
  Remain campaign 128, 131
  Twitter 'bombs' 74
Haugen, Frances 145
HERE 13
high-choice media environments 5, 100
highly-educated populations 10, 27, 28–29, 31, 51, 148
high-quality content 142
high-volume posting 70, 74, 75, 77, 113, 120
Hipp, John R. 50
homophily 50, 54, 57
Horwitz, Jeff 145
Howard, Philip 72, 73, 74, 76, 151
HTTP crawlers 122
human annotators/coders 71
human rights 29

hybrid media environments 49, 105
hybrid warfare 5
hyperactive accounts 82, 123
hyperpartisan content 93, 101, 118, 120, 129–135, 139, 140, 142, 143–144, 156, 159

I

identity affiliation 13, 29
  see also group identities
ideology
  clustering 52–55, 153
  ideological coordinates 33, 38–46
  polarization 30, 49–68, 72, 98, 105, 110, 139, 149, 150, 156
  value space 19, 40, 42–43, 148–149
image distribution 97, 100, 109, 119, 121, 140, 141
immigration 79–80, 151
impersonation of a third party 70
in-bubbles 13, 20, 58–68
  see also echo chambers
independent source attribution 108–109
individual rights 32, 37
inequality 28, 29, 45
influence operations 93–100, 111–112, 117–120, 125, 137–139, 158
information cascades 141
information cocoons 54
information diets 50–51
information diffusion 63–64, 73, 87, 93
information rectification 79, 111–112, 120
information warfare 52, 91–103, 105, 106, 118, 154–156
Inglehart, R. F. 19, 28–29, 30, 31, 34, 35, 36, 38, 40, 45, 80, 149
in-group favouritism 50, 56
Instagram 108, 151
intercoder reliability 39
international tweeters 14–15
Internet Archive 102, 141

Internet Research Agency
  (IRA) 74–75, 94–99, 100, 111,
  128, 151, 154
Iyengar, Shanto 50

J

Johnson, Boris 132, 135, 157
Johnston, Ron 57
journalistic quality 142, 145
Jowett, Garth S. 91–92

K

Kahan, Dan M. 52
Kollanyi, Bence 72, 73, 74, 76, 151
Kolmogorov-Smirnov 14, 63,
  64–66
Kremlin 74, 94, 95, 151
Kriesi, Hanspeter 30, 31, 32,
  35, 36
Kruspe, Anna 13

L

Labour Party 28, 30, 31, 35, 149
large language models (LLMs) 142
Leave campaign
  @vote_leave official
    account 83, 87
  anti-establishment 31
  bots 80, 100, 151
  Bristol troll farms 99–100
  deleted content 131, 132
  deleted tweets 123, 125–127, 157
  echo chambers 59, 61, 63, 67, 153
  geography of 34–36
  hashtags 12, 127, 128, 131, 132
  identification of messages
    from 13
  immigration as key Vote Leave
    message 79–80
  IRA Brexit campaign 98
  multiple hashtags 18
  nationalism 80, 150
  older voters 29
  rural areas 68
  tagging as 18
  'Take Back Control' 150

Leave voters
  demographics 37
  economism 37
  factors positively correlating
    with 42
  geographic distribution of 45
  nationalist sentiments 36–37
  populist sentiments 37
Leave.EU 98, 99, 100, 150
left-right politics 35
libcurl implementation 77
Liben-Nowell, David 50, 51
liberalism 35, 51, 56
linguistic signals 144
links, broken 101
Livingston, Steven 110, 111
local authority district
  matching 40, 42
location disguising 98
location field in user profiles 13
London 10, 22, 28, 32, 34, 44, 63,
  148–149, 150
longitude/latitude values 13–14
long-term nature of research 10
low education levels 29, 31, 36,
  80, 149
low-income communities 29, 36, 42
low-quality content 118, 121, 125,
  137, 138, 140–144, 155, 158

M

machine learning algorithms 20,
  34, 36–41, 42, 71, 146
mainstream media 5, 19, 98, 102,
  106, 142, 155
manipulation 4, 70, 93, 104–113,
  119, 137–139, 154, 158
manual coding 38–39, 40, 71, 97
manufacturing sector 42
mapping of users to electoral
  districts 7, 8, 21–23
marginalized groups 32
mass media 92–93, 106
Mastodon 107
Matthews, Miriam 111, 120
May, Theresa 131–132
McCreadie, Richard 122

INDEX

McGrory, C. 7–8
McNally, Naoise 147
mean partisan affiliation 18, 23
memes 100
mentions (@-)
 bots 70, 73
 distance travelled calculations 61
 echo chambers 60, 61
 identification of affiliation 18
 troll factories 96
Mercea, Dan 12, 20, 36, 74, 75, 93, 99, 106, 118, 120, 141
messaging apps 107
metadata 77–78, 79
Metaverse 107
Metaxas, Panagiotis 73
methodology 7–23
metropolitan elites 27, 28, 34, 44, 150
microtargeting 57–58, 104–113, 149, 153
middle classes 31
see also elites
Midlands 11, 22, 34, 45, 46, 62
migration 29, 32, 36
Miller, Nick 100
Miller, William L. 57
misinformation 70, 101, 104–113, 118, 139, 141–145, 155–156
misspellings 39
monetization of social media interactions 53
monthly active users (MAU) metric 8
Morozov, Evegeny 49
motivated reasoning biases 51
Mudde, Cas 31, 32, 36
multi-channel warfare 5
multi-dimensional engagement strategies 75
multimedia 82, 100–101, 140, 141, 155
Murthy, Dhiraj 72
Musk, Elon 3, 110
Mustafaraj, Eni 73

N

National Statistics Postcode Lookup 22
nationalism 34–46, 150
 Brexit Classifier algorithm 39–41
 in Brexit tweets 36–38, 42–46
 deleted content 127
 geographic distribution of tweets 38, 45
 Leave campaign 151
 nationalist parties 32
 nationalist populism 35–36
 nationalist rhetoric 28, 30, 31, 34, 38
 nationalist values 19, 20, 32–33
 Scotland 41, 45, 46
 strong until last few days of campaign 43
nativism 28, 32
neighbourhood effects 57
Nemorin, Selena 92
network effects 153
network externalities 52
networked publics 104–108
neutral (non-partisan) tweets 18, 19, 20, 59
news algorithms 144–147
news companies 146–147, 158
news cycles 112, 137, 158
niche subcommunities 146, 159
non-Brexit data set 20
non-UK located users 22
Norris, P. 19, 28–29, 30, 31, 34, 35, 36, 38, 40, 45, 80, 149
North of England 34, 45, 62
Northern Ireland 34, 149

O

Oard, Douglas W. 122, 123
O'Donnell, Victoria 91–92
older voters 29, 149
online activism and real-world events 49–50
online social networks 50–52
Onnela, Jukka-Pekka 50, 51

ONS (Office for National Statistics) 22
open APIs 109
open network infrastructures 105–106
openly partisan hashtags 128, 132, 157
orphaned data 120–123, 124, 137
OSLAUA 22
out-bubble communication 59–61, 62, 64–66
outgroup derogation 50
oversight mechanisms 6, 107, 109, 112–113

P

paper.li 82, 142
Pariser, Eli 54
Pattie, Charles 57
Paul, Christopher 111, 120
PCON11CD 22
peer effects 57–58
perceived threats 98
Peretti, Jonah 145
Pew Research Center 10
place_id 13
Pleroma 107
polarization 30, 49–68, 72, 98, 105, 110, 139, 149, 150, 156
political cleavages, maximizing 30
political realignment 27–33, 35, 151
populism
  bots 151
  Brexit Classifier algorithm 39–41
  in Brexit tweets 36–38, 42–46
  cultural backlash hypothesis 80
  definitions of 31–32
  deleted content 127
  hard to split from economism 41
  ideological clustering 20, 28, 29, 32–33
  IRA Brexit campaign 98
  in last few days of campaigning 43, 46
  nationalist populism 35–36
  regional distribution 45

postcode-level data 8, 13–14, 22
post-materialism 29
post-referendum activity 15, 99, 100, 123–124, 131–133, 155, 157
precision-recall trade-off 39–40, 71, 146
Prigozhin, Yevgeny 94, 95, 154–155
privacy 16–17, 107
Privacy Policy 17, 125
private profiles/accounts 16, 17, 123, 124, 129
'problematic content' 139–144
profiling technology 100, 108
progressive values 28–29, 34, 35, 149
prolific posting 74, 77
propaganda campaigns 91–95, 106, 117, 140, 154–155
protest activism 128, 132
provocative tweets 19
proximity in social networks 50
public APIs 109–110
Public Feed API 110
public record, altering of 119, 132, 136, 138, 156
public trace data 17
purging/downranking of disinformation content 111–112

Q

QAnon 111

R

racial attitudes 29
random sampling 15
Ratkiewicz, Jacob 72, 73
real-time research 1, 112, 118, 122, 137
recommender algorithms 144–147, 158–159
Reddit 75
regions 32, 34–36, 45, 148–149
'regrettable posts predictor' 122–123

INDEX

rehydration 15, 20, 121, 124, 128, 132
religion 29
Remain campaign
  deleted content 127, 131, 157
  deleted tweets 123, 125–127
  demographics 37
  echo chambers 59, 61, 63, 67, 153
  geography of 34–36
  hashtags 12, 18, 128, 131
  identification of messages from 13
  multiple hashtags 18
  tagging as 18
  urban-based users 67–68
RemainLeave variable 42–46
removal of accounts by Twitter 10, 87, 155
  see also deleted accounts
removal of posts from Twitter 10–11, 87, 111–112, 155
  see also deleted tweets
reply tweets 73
reporting tweets 112, 113
Republic of Ireland 35
REST API 12, 13, 15, 16, 23, 76, 123, 124, 131
retweets
  botnets 70, 80
  bots 72, 73, 75, 78, 153
  Brexit Bots 17
  in data sample 15
  deleted tweets 118–119, 124–125
  distance travelled calculations 61
  echo chambers 60, 61
  ethical considerations 17
  'low-quality content' 143
  retweet cascades 71, 78–79, 81, 82–87, 124–125, 151–152
  retweet networks 78, 96
  troll factories 96
reverse engineering 112, 118, 121, 156
reverse-geocoding 13–15, 21–23
rhetorical features 143, 144, 158
right to remember 136–139
rights 'of the people' 41

robustness checks 61
Rose, K. 7–8
Rosner, Bernard 16
Rovira Kaltwasser, Cristobal 31, 32
rumours 92, 98, 101
rural areas 68
Russia 74, 94, 95–99, 100, 111, 128, 151, 154

S

sanctioned archives 108, 119
scapegoating 98
Schou, Jannick 101, 106, 139
science 51
Scotland 28, 34, 38, 44, 45, 46, 149
Search 124
Seasonal Hybrid Extreme Studentized Deviate (S-H-ESD) algorithm 15–16, 132, 134
Segerberg, Alexandra 105
segmentation of users 108
self-government 37
self-referencing *bleus* 82, 102, 155
Selivanov, Dmitriy 20, 39
sensitive data 16
serial activists 75
shelf life of content 81, 138, 141, 142, 151, 158
  see also ephemerality
Sky News 102–103
slacktivism 49
'social bots' 72, 153
social identity theory 50
social infrastructure 106, 107
social networks 50–52, 56–58
Social Science One 108
social ties 50–51, 54, 56–57
socioeconomic realignment 28, 29, 35, 149
sockpuppets 70, 73–74, 99, 140, 150
source attribution 108
space-independent communities 51
spam policies 76
spillover effects 52, 57–58, 67, 153
Starbird, Kate 121

state actors 91–93, 94
Statista 10
stratification of social media 53
Streaming API 12, 13
stylistic devices as clues to low quality content 143, 144, 158
subcultures 96, 154
Sunstein, Cass R. 30, 52
superusers 67, 69–76
supervised high-volume posting 75, 151, 154
surrogate accounts 140
see also sockpuppets
surveillance 93, 107
suspension of accounts 10, 76–77, 101, 117, 120, 123–125, 131, 158

T

tabloid journalism 80, 82, 101, 102, 140, 146, 151
Tajfel, Henri 50
'Take Back Control' 37, 150
targeting material 100
see also microtargeting
team-managed Twitter accounts 74, 77, 154
Terms of Service 70, 109, 117, 120, 121, 124, 125, 141
text vectorization 20, 39, 40
TF-IDF method 40
thresholding approach to bot detection 76–79
tie formation 50, 54
time period of study 15
timestamps 79
top ten accounts 10
trade liberalization agreements 32, 35–36
traditional conservative values 30–31, 151
TREC Conference 122
Trending Topics 106, 145, 159
tribalism 52, 105, 156
troll factories/farms 94, 95–97, 99–103, 111, 154–156
trolls 16, 74–75, 152

Trump, Donald 35
turnout 42
tweet decay 120–129, 138, 156–157
tweet decay coefficient 15, 120, 131–132
Twitter
  continued use of name 3
  generous stance on data access 3, 6
  policy change re geocoordinates 13
  position in high-choice media environment 5
Twitter Compliance Firehose 123
Twitter Moderation Research Consortium 97
Twitter Streaming 16, 17
Twitterbots 16, 17

U

UK General Elections 121, 135, 156
UK users only, focus of work on 7
UKIP (UK Independence Party) 31, 42
Ukraine 98
unemployment 29, 37, 42
unique identifiers 79
universal commons, internet as 106
universalistic worldviews 32, 37
untrue statements 79
urban-based users 10, 67–68
URLS 100–101, 140–141
US District Court 95
US 'helicopter journalism' 102
US politics/elections 35, 72, 73, 93, 94, 95, 110, 142, 151
user data collection 13, 107
user rights 16
user-generated content 80, 82, 118, 139, 142–143, 151, 158
usernames 70

V

Vallis, Owen 16, 132
Varol, Onur 71, 76, 77
Vinhas, O. 120, 136

INDEX

viral content 78–79, 104
virtual reality 107
vocabulary-based vectorization 40
Vosoughi, Sorough 139

W

Wagner Group 94
Wales 34, 44, 45, 149
Walker, S. 101, 106, 119, 121–122, 137
ward-level data 21
web interface 78, 109, 123, 124
weblinks 101–102, 118, 140–141
Weedon, Jen 70, 110, 118, 140
welfare state 35
WhatsApp 5

white propaganda 93–94, 96, 98, 154
Woolley, Samuel C. 72, 74
working classes 28, 31, 34, 35, 36, 149

X

X, takeover of Twitter 3
xenophobia 31, 80, 151
Xu, Jun-Ming 122, 129, 138

Y

Yiannopoulos, Milo 75
YouTube 100, 144

Z

Zuckerberg, Mark 107